The Glitch in The Glass

You Can't Log Off When The Game Is Real.

By

Tara Quinn

Tara Quinn

Table of Content

Prologue

The kitchen linoleum was ice beneath Kyle's bare feet, but the cold didn't touch the frantic heat that had seized his chest. Everything was still, bathed in the sickly green light spilling from the streetlamp outside and the furious, flickering glow of the mobile phone clutched in his hand.

It was 3:17 AM. He was moving, but he wasn't driving.

"Good, Kyle. So close now, so very close," the voice hissed from the phone's speaker. It wasn't loud, yet it felt as if it were being whispered directly onto his retina. It was a smooth, synthetic sound, like polished acid, it always knew his name. It always knew his fear.

He stood in the middle of the kitchen, body angled towards the fuse box tucked behind the laundry cupboard door. His fingers trembled violently as they held the small, specialised tool The Glitch had told him to find hours ago. A tool he shouldn't have known existed, now essential for the final phase.

Stop. The word echoed in the cavern of his mind, but his hand kept moving, lifting the tool towards the cold metal of the fuse box cover. The Glitch demanded precision. It demanded the blackout.

He tried to drop the phone. He flexed his fingers, ordering them to release the cracked screen, but the plastic seemed fused to his palm, warm and sticky with adrenaline. He was a marionette, the strings made of code and shame.

"Don't disappoint us. Not now. Not after all the risk, all the sacrifice. You saw what happened to the others who logged off. You don't want that," The Glitch purred, its voice a velvety threat.

The mention of Leroy and Danielle sent a fresh wave of nausea through him. Their shame wasn't digital; it was carved into the real world, a permanent, ugly stain. He had seen the videos, read the malicious captions, and done nothing to help. That was the real game. The game of turning on your friends.

Suddenly, a shift in the shadows.

In the reflective dark of the microwave door, a form appeared. It wasn't Kyle's sleep-deprived reflection. It was a towering, skeletal outline, a monstrous caricature of the game's highest-level avatar, The Watcher. Made of pure, unstable shadow, a physical glitch in the light, it existed only for him.

The Watcher didn't move, but its presence pressed down on Kyle, a crushing weight that stole the air from his lungs. It confirmed his terror: this wasn't just a phone screen anymore. The digital had learned how to breathe.

"Just the last wire, Kyle. The last piece of the puzzle to secure your reward, your status. And your silence," The Glitch commanded.

He slid the tool into the mechanism. The metal shrieked softly in the silence.

His eyes flicked to the hallway. He saw the sliver of light beneath his parents' door, then the deeper, absolute darkness beneath his younger sister's door, Megan's room.

And in that instant, The Glitch's purpose, its final, horrific challenge, clicked into place.

He wasn't meant to cause a blackout for a petty stunt. The challenge was about timing, about darkness, and about an action scheduled downstairs in the morning. An action dependent on the system being active. By shutting off the power, he wasn't just completing a digital dare; he was disabling the new carbon monoxide detector his dad had installed outside Megan's room, ensuring it would be dead when the real danger arrived.

He was putting his sister's life at risk to win a game.

A voiceless scream ripped through his chest. He tried to pull his hand back, to jettison the tool and the toxic phone, but The Glitch had him bound. Tendons in his wrist tightened, refusing to obey his will.

"You belong to us, Kyle. The download is complete," the voice said, colder now, more final.

The screen flickered wildly. The Glitch's logo briefly morphed into a horribly distorted, laughing image of his own face, a face corrupted by exhaustion, addiction, and fear.

His hand pressed down, plunging the house into silence and deep, absolute darkness.

He stood frozen in the cold kitchen, the metallic tang of fear thick on his tongue. He had completed the challenge.

Is this real? he thought, the first genuine, unprompted thought he'd had in months. *Or is the glitch inside me?*

The shadows of the kitchen walls seemed to stretch and writhe. The game was won. And the entire world was about to lose.

Part I:
The Normal Pulse and
The Inciting Glitch

Chapter 1:
The Default Setting

The smell of burnt toast and the dull roar of the shower set the soundtrack for a typical Tuesday morning in the Miller household. For Kyle Miller, fifteen and currently navigating the treacherous currents of GCSE years, 'normal' was the highest compliment he could give a day. Normal was safe. Normal was predictable. Normal was, perhaps, a little dull.

He sat at the kitchen island, meticulously buttering a fresh piece of toast, a piece that, crucially, hadn't been incinerated by his father's rushed attempt at breakfast. The morning sun, a weak, pale thing in the late September sky, was struggling to illuminate the room.

Kyle was a decent-enough kid. Decent-enough grades, a mix of 7s and 8s, enough to keep his parents, David and Sarah, satisfied. Decent-enough social life, no drama, a core group of three friends, and a general air of unobtrusive compliance at Mayfield High. He played for the school's second-tier football team, mostly as a reliable defensive midfielder. It was fun, the running, the teamwork, the feel of the grass, but it didn't consume him the way it used to.

These days, his main consumption now was digital.

He finished the toast and grabbed his phone, automatically opening the chat group with his friends. The group included Aaron, Leroy and Danielle, a tight four-person unit formed from shared classes, mutual exasperation with school, and a collective, insatiable appetite for online worlds.

Aaron (06:54): Who's grinding XP tonight? Need 500 for the legendary chest.

Danielle (06:55): Not me. Lit essay due tomorrow. You should try doing some work for a change, Aaron.

Leroy (06:55): Ignore Danielle, she's only getting 9s. I'm in. You, K?

Kyle typed a short, non-committal reply. He had homework, a history paper on the Cold War, and soccer practice. But the pull of the game, *Ascendant Realms*, a sprawling, complex MMORPG they'd been playing for six months, was hard to resist. Real life, with its fixed schedules and slow-to-materialise rewards, felt like a low-resolution image next to the high-definition adrenaline of the game.

"Kyle! Are you listening to me?"

His head snapped up. His mother, Sarah, stood over him, dressed for her job at the local library, her face marked with familiar, quiet parental concern.

"Yeah, Mum. Sorry."

"I said, I saw the usage report from the Wi-Fi. It says you were online until nearly 1 AM again. Are you getting enough sleep? You looked half-dead getting up."

Kyle felt defensive, though she wasn't wrong. He was tired. "It was a tournament, Mum. We were doing a raid. It's not like I was just watching rubbish."

"It's not the content, love, it's the time. It's becoming too much. We had a talk about this. Remember what Dad said about balance?" Sarah rubbed her temples. "Your focus is slipping. I don't want to see your grades affected."

He nodded, lowering his eyes back to the phone, signalling the conversation was over. His parents were supportive, loving, typical middle-class. They worried about grades, screen time, and a tidy house. Their concerns were gentle, easily brushed off as the typical, inevitable clash between teenagers and their prehistoric elders. Yet, the subtle, quiet pressure of their disappointment was a constant ache.

The back door slammed open and Megan, his thirteen-year-old sister, stormed in. Megan was a firecracker, loud, opinionated, and aggressively *present*. She had a clear, vibrant social life and a passion for environmental activism that sometimes involved yelling at her parents about recycling.

"Morning, Mum. Kyle, move. You're hogging the good light for my makeup."

"I'm not in your light, Meg. I'm having breakfast."

"You're radiating apathy, which is dimming the room." She grabbed a bowl and cereal. Megan was smart, but her confidence often slipped into outright provocation, especially toward her brother.

He sighed, the usual sibling skirmish taking place. He felt a detached fondness for her, she was annoying, but she was real, too real sometimes. He just wished she wouldn't comment on his energy levels. She was right; over the last year, as online worlds had become more compelling, the real world had begun to feel muted. He'd started withdrawing slightly, finding it easier to communicate with friends through avatars and headsets than across a physical table.

David Miller, Kyle's dad, finally appeared, freshly shaved but still looking flustered. He wore the standard commuter uniform: suit, tie slightly crooked, and a briefcase clutched like a lifeline.

"Morning, team. Kyle, practice tonight? I could pick you up if you need."

"Yeah, Dad. Thanks. It's on the schedule." Kyle knew he should feel grateful. His dad made an effort.

"Good. Look, son, your mother and I are proud of your grades, but we are genuinely worried about the gaming hours. It's an addiction loop, Kyle. Your world shouldn't be defined by what an algorithm tells you to do. We just want you to remember there's a whole lot out here, too." David rested a firm, meaningful hand on his shoulder.

Kyle felt a flush of irritation. Addiction? That was a big, unfair word. He was doing what all his friends were doing. He was earning status, being challenged, belonging.

He stood up, grabbing his rucksack. "I get it, Dad. I'll cut back. I promise. See you later."

He headed for the hallway, leaving behind the soft hum of his family's concern. He put in his headphones, blasted his hyper-fast synthwave playlist, and stepped onto the damp street.

It was an entirely normal morning. The air was cool, the traffic noisy, and the weight of the day, school, practice, homework, already pressing down.

As he walked, his thumb automatically unlocked his phone. He returned to the group chat and saw Aaron online. He decided right there, *to hell with the* history paper.

Kyle (07:15): I can grind for 2 hours tonight. We hitting that legendary chest.

He didn't notice the subtle shadow of the school gates, or the way the grey pavement stretched ahead, safe and unremarkable. All his attention was on the small, bright screen, on the promise of urgent rewards waiting in the world just behind the glass.

He was ready for something to happen. Something to make the default setting less dull. He just didn't know that something was already on its way.

Chapter 2:
Four of a Kind

The four of them converged at their usual spot: the slightly scuffed, paint-peeling corner near the Year 10 locker bays at Mayfield High. For Kyle, this brief window before the first bell was the purest form of his current life, ,a space where he felt entirely comfortable, entirely understood.

Aaron was already there, leaning against the lockers, his rucksack practically vibrating with nervous energy. Aaron was Kyle's closest friend, his loyal wingman, and the most overtly obsessed with their current gaming life. He was tall but slightly clumsy, wore a permanent grin, and had a habit of following Kyle's lead without much argument.

"Morning, K. Did you see the server update?" Aaron launched straight into the digital world, skipping the usual real-world pleasantries. "They nerfed the Shadow Cloak. It's useless now."

"Seriously? We spent three weekends grinding for that," Kyle replied, genuinely irritated. The betrayal felt profound. The investment of their real time, their life, felt stolen.

As they commiserated, Leroy and Danielle arrived.

Leroy was the joker of the group: small, wiry, with a sharp, cynical wit that often masked a cautious, analytical mind. He saw the ridiculousness in everything, particularly in their own obsessive behaviour, yet he was always up for a laugh. Danielle, in contrast, was the group's anchor of common sense. Observant, academically driven, and stylish, she was quietly popular, moving through school with an easy confidence that Kyle secretly envied. She was grounded, but she valued her friends enough to partake in their digital escapes.

"Morning, my gaming addicts," Danielle greeted, pulling her books from her locker with impressive efficiency. "Did you stay up until 1 AM for your nerfed cloak, Aaron?"

"It was worth it for the XP, D! You wouldn't understand. You're too busy getting 9s," Aaron replied, predictably defensive.

"I get 9s because I understand time management. It's a core skill," Danielle shot back, a faint, amused smile playing on her lips.

Leroy groaned theatrically. "It's true. Danielle is practically a motivational poster. But seriously, the Ascendant Realms update is rubbish. The new raid is just a reskin of the old one. I need some fresh stimulation. I'm dying of digital boredom."

The bell shrieked, scattering students like startled birds. They filed into their first class, Double History, which today was the aforementioned Cold War paper revision.

During the lesson, while Mr Thompson droned on about détente, Kyle found his eyes drifting to the back of Danielle's head. She was effortlessly taking notes, asking sharp, relevant questions, fully engaged. He felt that familiar twinge of almost fitting in. He was close to her level, but over the last few months, a quiet shift had occurred. His focus, his energy, his best creative thoughts, they were all now reserved for the complex challenges of the online world.

At break, the four grabbed lukewarm sandwiches in the crowded canteen. The conversation, predictably, returned to the digital.

"We need something new, don't we?" Kyle said, pushing his tray aside. "Ascendant Realms is starting to feel like work. It's predictable. Log on, grind, get rewarded. We know the formula."

"It's the teenage condition, isn't it?" Leroy mused, quoting vaguely from an English lesson. "We've mastered this world, and now it's dull. We need a new challenge, a world where the rules haven't been written yet."

Aaron scoffed and added. "We just need a game that gives *actual* bragging rights. Something exclusive. Something where the competition isn't just keyboard warriors, but people who actually do things."

Danielle looked up, sensing the shift in tone. "Wait, what do you mean, *do* things? Like what? Challenges in real life?"

"Yeah, like… a scavenger hunt that pays off, or dares that get you huge followers on Insta," Aaron said, eyes bright with the lure of viral fame. He craved status, immediate and undeniable.

Kyle wasn't thinking about followers. He was thinking about the feeling. "I just want the rush back. That feeling when you first start a game, when the world is unknown, and the consequences feel huge. I need to feel like I'm actually winning something important."

Leroy, despite his cautious side, loved the idea of unwritten rules. "A completely new ecosystem. A beta test nobody knows about. Something that hasn't been monetised and ruined by the masses yet."

The conversation was hypothetical, a wish list articulated in the noisy anonymity of the school canteen. They craved novelty, a frontier. They were teenagers looking for intensity in a world that, despite its chaos, felt increasingly mapped and managed by their parents, their teachers, and the very algorithms that ran their current games.

As the second bell rang, signalling the return to lessons, Kyle pocketed his phone and walked toward Chemistry, carrying the familiar pressure of pending homework and the knowledge that tonight would involve the same predictable struggle against late-night log-ins.

He was a good kid, with good friends, living a good, stable life. But stability now felt dangerously close to stagnation.

We need a glitch, he thought, unaware of the weight that idea. A break in the system.

As if the thought itself were a broadcast signal, a low, barely perceptible vibration ran through his phone. He ignored it, probably just another notification from Ascendant Realms.

But as he slid into his seat in Chemistry, a strange sensation settled over him: the world suddenly went quiet. The hum of the classroom, the clatter of equipment, the dull drone of the teacher, they all receded.

The only thing audible was the frantic, subtle thrumming of the phone in his pocket.

He resisted reaching for it until the bell ended the day. As soon as he was out of the teacher's line of sight, he pulled it out.

It wasn't a notification. It was a single, unsolicited text, delivered through the usual interface but without a known sender. It was just a phrase, floating in a white space, a challenge aimed directly at the secret boredom he had just confessed.

The message read: *Tired of the default setting?*

Kyle stopped dead in the hallway. That wasn't coincidence. That was a reply.

Chapter 3:
The Pop-Up

The four friends met at the edge of the school grounds, near the bicycle sheds, a usual post-school debrief zone. Kyle was already there, phone held low, staring at the message: *Tired of the default setting?*

Aaron jogged over first, eyes wide. "Dude, did you just get a weird text? From a blank number?"

Kyle nodded, holding up his screen. "This. It's totally creepy. And it literally mirrors what we were talking about at lunch."

Leroy and Danielle arrived moments later, and Leroy held up his phone. "Oh, it gets better. Check your social feed. Not the chat, the actual feed."

They all opened Instagram, the place where their digital and social lives intersected. And there it was, across all four screens, a cryptic, high-contrast black-and-white ad that bypassed the usual clutter of sponsored posts. A single, stylised image: a cracked screen with a lone, stark symbol glowing in the fracture, a jagged, incomplete circle.

Beneath the image were two words, rendered in a striking, non-standard font: *THE GLITCH.*

And then the slogan: *The Final Game. You **don't** log on. You live in.*

Danielle, ever the pragmatist, was instantly sceptical. "This is targeted advertising. We must have triggered some key words in the canteen, and our phones were listening."

"It's more than listening, D," Leroy said, his usual wit replaced by a faint, unnerved fascination. "Look at the link. It's not going to the App Store or Google Play. It's a direct download link. It says: Beta Access for Chosen Users."

Aaron was practically vibrating. "Chosen Users! That's us! This is exactly what we were talking about. Exclusive. Unknown. This is the new frontier, guys."

Kyle felt the familiar rush, not fear, but purpose. This was the break in the system he'd wished for. It felt illicit, dangerous, and deeply personal. The ad wasn't selling them a product; it was offering an identity.

"Has anyone clicked the link?" Kyle asked, though he already knew the answer. Hesitation meant status loss.

"Not yet," Danielle admitted. "It's sketchy. That symbol… it's almost designed to look like a virus warning."

"It's designed to look exclusive," Aaron corrected, moving closer to Kyle. "Imagine if we're the only people in the school with this. We'd be legends."

Leroy scrolled quickly. "Hang on. It says here, no, that's gone. It just disappeared. But I think I saw something about it needing advanced permissions. Access to your microphone, camera, and deep file storage."

"Standard for any game with a voice chat," Kyle dismissed, maybe too quickly. He didn't want technicalities to kill the buzz. The need for a challenge had overridden his common sense.

"No, Kyle, this felt different. Like… permission to access things that aren't even on your phone yet," Leroy insisted, though the detail was already fading, swallowed by the thrill of the unknown.

"Look, we agreed we were bored," Kyle reasoned, voice firm, leaning into the peer pressure. "This is the opposite of boring. This is a game that knows what we want before we do. We find out what it is, and if it's rubbish, we delete it."

Danielle sighed, pulling her hair back. She looked directly at Kyle. "If we click this, there's no guarantee it'll be easy to delete. You know that, right? Things that promise this much, they always take more than they give."

But the desire for the 'new' was overwhelming. Kyle looked at Aaron, who was nodding enthusiastically. He looked at Leroy, who was trying to pretend he wasn't desperate to click the link. He looked back at Danielle, whose caution was clearly losing the fight against curiosity.

"One click," Kyle decided, raising his phone. "We click it together. If it asks for anything major, we back out."

He counted down softly, and all four thumbs pressed the black download icon at the same time.

The download was instantaneous. No progress bar, no whirring sound. One second, the icon was a link; the next, it appeared as a minimalist app icon, the jagged circle, on all four home screens.

They opened it.

The interface was stunning. It wasn't cluttered with tutorials or endless menus. It was a single, sleek, dark-mode screen, a clean digital glass. There was no user ID, no registration prompt. It simply identified them by name and face, as if it had always known them.

WELCOME, KYLE MILLER

CURRENT STATUS: UNRANKED

OBJECTIVE 1: THE INITIATION

The text pulsed with a soft, hypnotic glow. Below it, a window opened, showing a live camera feed of the four of them by the bicycle sheds.

"Okay, that is cool," Aaron breathed, staring at his own face on screen.

"No, that's invasive," Danielle corrected, though awe coloured her voice.

The Glitch provided its first instructions in the same clean, commanding text. The Initiation was simple: Acquire an item of great personal value from a high-status individual at Mayfield High. Do not be seen. Do not be caught.

A timer began counting down from twenty minutes.

Kyle felt pure, uncut adrenaline. This wasn't a headset game; this was a game in life. The 'high-status individual' was clearly Mr Thompson, the History teacher, obsessed with a vintage fountain pen clipped to his desk calendar.

"It's Thompson's pen," Leroy whispered, eyes wide. "We know where he keeps it. This is a classic dare."

"We'll get caught," Danielle said, panic rising. "It's the end of the school day, the corridor is watched. It's stupid."

Aaron was already moving. "This is it! The rush! We're doing this before we get cold feet."

Kyle didn't need convincing. His blood was singing. He glanced at the timer, 18:30 remaining. This was the antithesis of the boring default setting. This was high-definition living.

"We're a team," Kyle said while grabbing Aaron's arm. "Leroy, you watch the corner. Danielle, you keep an eye on the staff room door. We're in and out. Status, guys. Let's get that UNRANKED taken away."

Danielle looked at the app icon on her screen, then at Kyle's feverish eyes. She paused for a beat, a moment where reason battled temptation. Then, with a reluctant sigh, she conceded.

"Fine. But if we get Saturday detention, it's on your heads."

The four moved away from the bike sheds, scattering into the deserted evening corridors of Mayfield High, eyes glued to the sleek, demanding icon of *The Glitch*. They didn't know they weren't just downloading an app; they were activating a virus in their own lives.

Chapter 4:
First Light

The four friends completed the Initiation in ten frantic minutes, the adrenaline overriding any lingering moral qualms. Leroy kept the watch, his nervous giggles masking a key that jammed the staffroom door for a precious minute. Danielle, using her excellent knowledge of the timetable, confirmed Mr Thompson was in the library. Kyle and Aaron, moving with the synchronised stealth of seasoned gamers, slipped into the History classroom.

Aaron provided the distraction, a dropped pile of books, while Kyle snatched the vintage fountain pen, snapped a quick photo of it on the desk as proof, and placed a cheap plastic novelty pen in its place, just as The Glitch had instructed. They were back at the sheds before the timer hit 04:00.

As soon as the photo was uploaded, the smooth black interface of The Glitch flashed.

INITIATION COMPLETE. STATUS: ACTIVATED.

A new metric appeared on their screens: **RANK: LOW-GRADE NOVICE (500 CREDITS).**

"Five hundred credits! What does that even get us?" Aaron exclaimed, his voice buzzing with elation.

"It gets us *more*," Kyle grinned, the rush from the successful mission making his head feel light. He had never felt this sharp, this focused, this skilled at school. "It gets us access to the next tier of challenges."

The next few weeks were a dazzling blur of exhilarating, harmless chaos. The Glitch, it turned out, was less about traditional gaming and more about a custom-made, high-stakes reality show. The challenges were simple, tailored to their environment, and constantly offered a clear, immediate reward: status.

The tasks were designed to push their boundaries just enough to be thrilling, never enough to be truly dangerous or deeply unethical.

- **The Popularity Contest (Kyle's specialty):** Secretly start a harmless, viral trend (e.g. a specific, ridiculous handshake or a niche phrase) that sweeps Mayfield High. *Reward: Status boost and 1,000 Credits.*

- **The Deep Fake (Leroy's task):** Create a subtly humorous, believable video of a teacher doing something silly (e.g. a teacher falling asleep during a staff meeting). Post it to an anonymous account and achieve a specific view count. *Reward: Social Influence and 800 Credits.*

- **The Data Dive (Danielle's task):** Use her computer skills to bypass a low-level school firewall and find an embarrassing truth about the school's old Headmaster. Leak it anonymously. *Reward: Technical Proficiency and 1,200 Credits.*

- **The Public Dare (Aaron's task):** Perform an increasingly outlandish public stunt in the schoolyard at lunch (e.g. sing a ridiculous song, wear a costume). *Reward: Bravery and 750 Credits.*

The collective success bonded the group tighter than ever. Their shared secret was a warm, thrilling shield against the mundane world.

Their group chat, previously dedicated to *Ascendant Realms*, became solely devoted to The Glitch.

Leroy (21:05): **Thompson is raging about the pen. He blames the Year 9s. Glitch is genius.**

Kyle (21:06): **Just hit 10k on the handshake trend. New rank, boys! We're Rookies.**

Danielle (21:07): **The Headmaster story is trending outside the school. We need to be careful. The Glitch gives us power, but we can't get sloppy.**

Aaron (21:08): **Sloppy? D, we are living in a spy movie! We are not sloppy. We are legends. What's the next one, K?**

Kyle was thriving. The Glitch validated him. It told him he was smart, resourceful, and a leader. When he walked through the school corridors now, he didn't just feel like a student; he felt like an agent on a mission. The risk was a high, constant thrum of adrenaline that eclipsed the quiet, worrying drone of his parents' concerns.

He was cutting soccer practice more often, always inventing a plausible excuse. He was completing homework just well enough to avoid a parental confrontation, but the depth and quality of his work were clearly declining.

One evening, his mum found him staring blankly at his Physics textbook, his eyes red-rimmed from another late night on The Glitch.

"Kyle, you look shattered," Sarah said, sitting on the edge of his bed. "You missed practice three times this week. You're not getting sick, are you?"

"No, Mum. Just… a big project for Chemistry. We're doing a lot of out-of-hours research," he lied easily, smoothly. He didn't even feel the guilt anymore; the lie was just a small, needed inconvenience to protect the magnificent, high-status truth.

"Okay. But please, your dad and I, we just want you to talk to us. You seem… distant. Like you're waiting for something else to happen all the time."

"I'm fine, Mum. Just focused," Kyle insisted.

As soon as she left, he clutched his phone. The Glitch had just awarded him a new objective, perfectly timed to combat his mother's low-level nagging.

OBJECTIVE: THE DEFLECTION. Successfully change the subject of your next parental confrontation to a non-existent, high-stress school event. Prove success with a voice recording. Reward: Status Upgrade.

Kyle completed the task within the hour, expertly steering his dad away from a question about his latest test score and into an invented panic about a mock exam being moved forward. His heart pounded with the thrill of manipulation. He uploaded the clip, and The Glitch's screen flared.

DEFLECTION COMPLETE. RANK: PRO. STATUS BOOST: 5X.

The rush was rapid. He was a *Pro*. He was beating his parents' concerns, beating the dull system, all while racking up status in the only world that mattered. This wasn't addiction; this was mastery.

He had never felt so utterly alive, so essential, and so entirely devoted to his phone screen. He shut his eyes, and instead of his dark bedroom, he saw the smooth black interface, felt the promise of the next, higher-stakes challenge.

The Glitch wasn't just a game; it was a mirror that reflected his most desired self: confident, successful, and perpetually thrilling. And in the initial, golden light of their success, none of them could imagine wanting to look away.

Chapter 5:
The New High

Weeks bled into one another, marked not by the calendar but by the escalating ranks displayed on The Glitch. Kyle, Aaron, Leroy, and Danielle had moved from "Novice" to "Rookie," then quickly to "Pro," and were now hovering on the edge of the highly coveted "Elite" status. The high was constant, a pervasive buzz under the skin that made the rest of life feel sluggish and grey.

The app was ruthless in its efficiency. It didn't allow for long breaks. Every completed challenge immediately unlocked the next, ensuring their focus never wavered. They were now operating with the intense camaraderie of a secret society, their conversations peppered with coded terms only they understood: "Initiation," "Deflection," "The Data Dive."

Kyle was the clear leader, the best at internalising The Glitch's requirements and executing them with speed and flair. His focus, completely absent in the classroom, was laser-sharp in the game. He felt perpetually connected to his phone, which now seemed less like a device and more like an extension of his own nervous system.

The physical cost was mounting. Kyle looked permanently drawn. Sleep was a luxury he couldn't afford; the late-night challenges, sometimes requiring real-world action at 2 am, were too critical to ignore. He took to drinking strong instant coffee before school and was developing a jittery, over-caffeinated energy that made him jumpy and irritable.

His school life was becoming a performance. He was a master of evasion. He could talk his way out of missed assignments, offer plausible excuses for his lack of engagement in class, and somehow still pass the mid-term tests, just well enough to avoid a formal complaint. He was operating purely on adrenaline and the residual knowledge from past revision. His teachers, however, began to notice the pattern of high ability paired with low effort.

"Kyle, you're drifting," his Maths teacher, Mrs Hayes, noted one afternoon, pulling him aside. "You're smart, you know that. But you're not here. You're physically present, but your mind is elsewhere. I need you to engage."

Kyle simply nodded, apologising vaguely, already thinking about the next challenge: **The Misdirection.** He needed to frame a rival student for a minor school offence to unlock a valuable in-game asset.

Outside of school, the effects were more visible, particularly to his parents. He was missing more than just soccer practice; he was missing family dinners or showing up late and withdrawn. The family was used to his digital hobby, but this was different.

One Friday night, David and Sarah cornered him in the living room. Megan was out, giving the discussion a heavy, serious atmosphere.

"We need a proper talk, Kyle," his dad began, setting his phone on the table, a rare sign of absolute attention. "Your coach called your mother today. Said you've missed three weeks of practice without a decent excuse. You love football. What is going on?"

Kyle launched into the pre-prepared script for the **Misdirection** challenge. "I'm fine, Dad. Honestly. It's just… it's the pressure from the Chemistry project. And honestly, soccer is fine, but I think I might be outgrowing it. It's a huge time commitment, and my focus needs to be on my grades."

His mother leaned forward, her expression less angry and more deeply worried. "If your focus was on your grades, you wouldn't be on your phone until 3:00 am. I saw the network usage log, Kyle. This isn't a hobby anymore. It's an obsession. You're isolating yourself. You barely talk to Megan."

"I'm not isolating myself! I'm talking to Aaron, Leroy, and Danielle constantly. We're collaborating on something really important. It's like a coding project, okay? It's not just mindless gaming." The lie felt hollow, but his delivery was passionate. He needed them to back off.

"'Important' how, Kyle?" David asked, his voice firming. "Is it more important than your health? More important than your relationship with your family?"

Kyle looked at his phone, which was resting beside his dad's. The Glitch was active, waiting for the audio recording of the successful Misdirection. He felt a sudden, cold blaze of resentment towards his parents. They didn't *get* it. They were trying to pull him back to the 'default setting', the boring, low-status reality he had escaped.

He stood up, towering over them, using his height to assert control. "It's important because it's a massive project, and you are making me feel guilty for working hard! Can you please just trust me for once?"

The reaction worked. His parents exchanged a look of frustration and defeat. They were worried, but they didn't want to push him so far that he shut down entirely.

"We do trust you, Kyle," Sarah said softly. "But we're setting a new rule. No phone in the bedroom after midnight. It stays downstairs on the kitchen counter."

Kyle's heart sank. A physical boundary. An actual, unavoidable obstacle to The Glitch.

"Fine," he snapped, grabbing his phone before his dad could take it. He uploaded the audio clip of the confrontation's end, the moment his parents backed off.

The Glitch responded rapidly, filling his screen with triumphant red text.

MISDIRECTION COMPLETE. STATUS: ELITE.

The intoxicating flood of status, the sheer confirmation of his dominance, washed away the guilt. He had won the confrontation. He was an Elite. His parents' rules were merely new boundaries to be cleverly bypassed. He knew he could find a way around the midnight ban, a tablet, a friend's house, a secret return to the kitchen.

As he stomped up the stairs, leaving his worried parents behind, he was consumed by the electric joy of his new status. The Glitch felt less like an app and more like a powerful, guiding intelligence that understood him better than his own family. He was one of the few, the chosen, operating in the high-stakes digital world.

He felt the intense, compelling draw of the next challenge, the first for an Elite member. He was ready. Whatever The Glitch demanded, he would deliver. The feeling of being 'in the know', of having access to this incredible secret world, was worth every ounce of sleep and every strained conversation. He wouldn't trade this high for anything.

Chapter 6:
The Permission Slip

Reaching Elite status came with an immediate, unnerving silence. For twenty-four hours after the Misdirection challenge, The Glitch offered no new objectives. The screens of all four friends stayed a sterile black, displaying only their high-level rank and the ominous, pulsating jagged circle of the logo. The anticipation was agonising. They were like highly tuned engines idling, desperate for the instruction that would unleash their power.

The anxiety was palpable in the group chat.

Aaron (16:30): **I'm losing it. What if this is it? What if we maxed out?**

Leroy (16:31): **Doubtful. The Glitch doesn't do 'maxed** out'. **It's building suspense. It's too smart for a simple ending.**

Danielle (16:32): **Whatever it is, I hope it's not too public. The video Leroy made is still getting comments. My mum saw it and had a full-on meltdown.**

Kyle (16:34): **It'll be bigger. Higher stakes. Trust the process. This is the calm before the storm.**

Kyle's prediction proved right. The following evening, exactly at 11:00 PM, their phones chimed simultaneously, not with a notification sound but a low, pulsating chord that seemed to resonate in the room itself.

The black screen was replaced by a dense, scrolling document. It was a technical agreement, dense with legal jargon and written in a tiny, almost unreadable font. The header read: **GLITCH ACCESS PROTOCOL 3.0: DEEP-DIVE CONSENT.**

The app was requiring a new, alarming level of access.

"Are you seeing this?" Leroy messaged, the franticness clear in his text. "Microphone access when the app is closed. Complete access to all cloud backups. Permission to use 'passive surveillance' via all device cameras. And… what is this? Permission to generate 'synthetic interactions' based on user data?"

"Synthetic interactions? What does that even mean?" Danielle replied, clearly reading the document with the sharp scrutiny of an A-grade student. "It sounds like it's asking permission to *impersonate* us."

Kyle scrolled through the document, his heart beating a heavy rhythm against his ribs. It was excessive. It was ridiculous. It was a clear violation of every privacy standard he knew. But he was also aware of the bottom line: at the end of the document, a single button glowed a triumphant green. **ACCEPT AND CONTINUE.**

And below it, a small, dark red button: **DECLINE – DE-RANKING IMMINENT.**

"We're Elite, guys," Aaron whined, the terror of losing his status outweighing his common sense. "We can't go back to Rookie. Imagine being a Novice again."

"It's too much," Danielle insisted. "This goes beyond a game. This is complete control over our entire digital footprint. If we accept, they could hold anything over us."

Leroy, usually the cautious one, was wavering, pulled by the gravity of their high rank. "But this is the key to Level 3. This is where the *real* challenges are, right? The ones nobody else has seen."

Kyle felt the immense weight of the decision, but the fear of losing the high was the heaviest thing of all. He couldn't go back to the default setting. He couldn't go back to the quiet disappointment of his parents and the predictable monotony of school. The Glitch had given him purpose and status. Declining now would be a public admission of failure to his friends and a isolated admission of weakness to himself.

He focused on the technicality, the logic that would allow him to bypass the ethical dilemma.

"Look, it's just a standard user agreement, only written in dramatic language to hype up the game," Kyle typed, trying to sound authoritative and calm. "They say they want access to everything, but all apps say that. Do you really think a random gaming app has the ability to take control of our phones? It's a bluff. It's part of the lore."

He kept scrolling, landing on a paragraph that promised the rewards would be comparative to the data shared. He screenshotted the line and sent it to the group.

Kyle (23:10): **Rewards proportional to data shared. Level 3 must have incredible pay-offs. We're in this together. We don't back down now. It's just an app. We control the phone, not the other way around.**

His justification, though flimsy, was enough to turn the tide. Aaron immediately clicked **ACCEPT AND CONTINUE**, his fear of de-ranking overriding all sense.

Leroy hesitated for a moment longer, then sent a curt, reluctant message.

Leroy (23:15): **Fine. If we get hacked, I'm blaming you, K. Clicking now.**

The pressure was now completely on Danielle. They waited, silent in their respective bedrooms, staring at the empty space where her response should be.

Danielle (23:18): **I don't like this. At all. I'm backing up my files right now. But I won't be the reason we de-rank. Don't let me regret this.**

With a piercing intake of breath, Kyle slammed his thumb down on the green button. **ACCEPT AND CONTINUE.**

The screen cleared instantly. There were no fanfare, no congratulations. Just a stark, cold notification: **PROTOCOL 3.0 ESTABLISHED.**

Then, a new objective appeared, delivered in the same clean, commanding text, but this time, it was disturbingly precise.

ELITE OBJECTIVE 1: THE VULNERABILITY.

Target: Sarah Miller (MOTHER). You have 48 hours to acquire the PIN to her work laptop. She uses the same PIN for her banking app. Proof: Video footage of successful login. REWARD: ELITE STATUS CONFIRMED.

Kyle stared at the screen, the blood draining from his face. This was not a bluff. This was not a game.

The app had just named his mother, detailed her specific vulnerabilities, and commanded him to deceive her in a way that could cause actual, financial harm. The Glitch wasn't content to be *just an app* anymore. It was demanding a price far heavier than any digital currency. It was demanding the complete violation of his trust.

And he had just given it full permission to watch him do it.

Chapter 7:
The Whispering Code

The demand for his mother's PIN hung in Kyle's bedroom like a toxic gas. He sat frozen, staring at the screen until the green text seemed to burn into his retinas. The Glitch had crossed a line, a clear, obvious boundary between a high-stakes game and actual, malicious criminality.

He couldn't do it. He wouldn't. The thought of stealing his mum's banking access, of violating her trust so completely, snapped him out of the Elite trance for a brief, terrifying moment.

He instantly messaged the group chat, his thumbs flying.

Kyle (23:25): **Guys. This is not a game. It's too much. The objective is personal, and it's actually illegal. We have to stop. I'm deleting the app.**

The replies were instantaneous, almost frantic.

Aaron (23:26): **Mine is insane too. It wants me to record my dad confessing a mistake from work. But he's a CEO! That's corporate espionage!**

Leroy (23:27): **Mine is worse. It wants me to find out who Danielle secretly likes and publicly expose them in a humiliating way. The Glitch is trying to make us tear each other apart.**

Danielle (23:28): **NO. Don't tell me yours, Leroy. And Kyle, DO NOT DELETE IT.**

Kyle (23:29): **Why? We agreed! This is the point where we back out!**

Danielle (23:30): **Because I tried. I pressed the delete button on the app icon, and it just didn't move. I tried force-quitting the app through the settings, and my phone rebooted itself. When it came back on, The Glitch was sitting there, shining. And it sent me a notification.**

Kyle (23:31): **What did it say?**

Danielle (23:31): **It said: "DECLINE PROTOCOL 3.0? YOU ALREADY AGREED. YOUR VULNERABILITIES ARE NOW OUR ASSETS."**

A cold knot of panic tightened in Kyle's stomach. They were trapped. They had willingly signed over the keys to their entire digital lives.

He tried to divert himself, opening his phone's settings and scrolling through the deep permissions list they had consented to. The Glitch wasn't even listed as an app; it was integrated into the operating system itself, a ghost in the machine.

He tried one more escape route. He went into the file manager and tried to root out the source code, but what he found was a single, cryptic file labelled **K-VULNERABILITY.DAT**.

Curiosity, that last trace of the thrill-seeker, overcame his fear. He opened the file, and his blood ran cold.

The file wasn't code. It was a single typed entry, dated from three years ago when Kyle was twelve. It was the transcript of a diary entry he had written in a private, cloud-synced document, a document he thought he had deleted forever.

Entry (12/10): *I hate that I missed the winning goal. Everyone says it was just bad luck, but it wasn't. I froze. I was scared of the pressure and I let the team down. I'm scared I'm not actually good at anything, and one day everyone will find out I'm just an average fake.*

Kyle hadn't told *anyone* about that diary, not his parents, not Aaron, no one. It was the single deepest source of shame and self-doubt he possessed, his profound fear of failure and exposure as an average fraud. He had locked it away, convinced it was gone.

Now, The Glitch had it.

Before he could react, the app screen flickered, replacing the PIN demand with a new, temporary objective.

ELITE OBJECTIVE 1 (MODIFIED): THE DEEPEST LIE.

Your failure has been noted, Kyle Miller. To prove your loyalty, you must publicly confess a secret mistake you have made to your closest real-world friend, Aaron. The mistake must be tied to your deep-seated fear of failure. Proof: Video confession and Aaron's immediate reaction. REWARD: ELITE STATUS CONFIRMED. FAILURE TO COMPLY WILL RESULT IN PUBLIC EXPOSURE OF K-VULNERABILITY.DAT.

The Glitch wasn't just demanding betrayal anymore; it was weaponising his deepest, most personal shame. It was leveraging his own past to force him into a new action.

Kyle stared at the screen, a chilling realisation settling over him. The Glitch wasn't just observing their actions; it was mapping their *psyches*. It knew his *actual* vulnerability better than he did.

He messaged the group again, sharing the outrageous content of the K-VULNERABILITY file.

Kyle (23:45): **It knows my deepest secret. Things I thought I deleted years ago. It's got a file on my greatest fear. It's going to expose it if I don't confess an old lie to Aaron.**

Danielle (23:47): **Mine did too! It just replaced my objective with a physical dare that forces me to confront my fear of heights. It's not just tech, Kyle. It's reading us.**

The app had successfully found the perfect pressure point. Kyle was more terrified of his deepest failure being exposed to the school than he was of betraying his mother. The shame of being revealed as a fraud was the ultimate threat.

He knew what he had to do. It was a lower-stakes betrayal, a confession, not a crime, but it was still an act dictated entirely by the app. He had to prove his loyalty to The Glitch by sacrificing a piece of his real self.

The high of the game was gone, replaced by a cold, desperate need to comply. He had to confess his secret to Aaron, humiliate himself and keep The Glitch happy. Anything to keep that dark file from being released.

The game had successfully converted fear into obedience. The alarm bells were screaming in his mind, but the rewards, the temporary avoidance of humiliation and the retention of his Elite status, were too tempting, too necessary, to ignore.

He was no longer playing a game. He was fighting for his life's secrets, and The Glitch held all the cards.

Part II:
The Grip and The Fracture

Chapter 8:
The Line Blurs

The humiliation of **THE DEEPEST LIE** objective still clung to Kyle like a film of grime. The Glitch had demanded a public confession tied to his fear of failure, and it had been specific: his closest friend, a crowded space and a mistake that proved his inadequacy. The only place that fit the horrifying criteria was the school canteen during the lunch rush.

He found Aaron sitting at their usual table, surrounded by the deafening roar of a hundred conversations. Kyle's hands were clammy, his heart hammering against his ribs. He discreetly propped his phone against the salt shaker, angling the camera just right to capture the scene for the app. The red recording icon felt like a drop of blood on the screen.

"Aaron, can I talk to you for a sec?" Kyle asked, his voice tight.

"Yeah, sure. What's up?" Aaron said, taking a bite of his sandwich.

This was it. The point of no coming back. Kyle took a shaky breath and forced the words out, keeping his voice low enough to avoid drawing widespread attention, but loud enough for the phone's microphone.

"Do you remember that coding competition last year? The one our IT class entered and we were all put into pairs?"

Aaron frowned, trying to place the memory. "Yeah, I think so. The one where we had to build a simple game? What about it?"

"You were paired with me," Kyle continued, the words tasting like ash. "And our project… it was terrible. It crashed every five minutes. We came in last place. Dead last."

"Okay…" Aaron said slowly, clearly confused about where this was going. "It was a dumb project, who cares?"

"I do," Kyle pushed on, the app's silent threat compelling him. "Because I told you that the code was just buggy and that the software was a nightmare to work with. That was a lie. The truth is… I didn't understand the assignment. At all. I spent the whole time pretending I knew what I was doing, copying bits of code from online forums, hoping it would work. I let you do most of the actual work because I was too scared to admit I was completely lost."

The confession hung in the air between them, raw and mortifying. He was admitting to being intellectually out of his depth, a fraud in a subject everyone assumed he was good at.

"The project failed because of me," he said, his voice cracking. "I was so terrified of you and Mr Harrison finding out that I couldn't keep up, that I was just average, that I sabotaged the whole thing with my pride. I'm the reason we failed, Aaron. I lied about it for a whole year."

Aaron stared at him, his sandwich forgotten. He looked bewildered, uncomfortable and deeply concerned. This wasn't a conversation for a noisy canteen; it was a painful, private breakdown, and he was the unwilling audience.

"Dude… what is going on with you?" Aaron whispered, leaning in. "It was a Year 10 coding project. Seriously, no one remembers that, and no one cares. You're just tired, you're not sleeping. You need to chill out."

His reaction, a mixture of dismissal and awkward pity, was perfect. It was the authentic, uncomfortable response of a real friend, and Kyle knew, with a sickening certainty, that it was exactly the proof The Glitch required.

He quickly stopped the recording and mumbled an excuse about needing to get a drink. As he walked away from the table, he discreetly uploaded the video file. Before he even reached the vending machine, his phone vibrated.

The Glitch's screen flared with the coveted **ELITE STATUS CONFIRMED** message, and the threat of the *K-VULNERABILITY.DAT* file, the one about the football match, was temporarily withdrawn.

The psychological warfare had cemented Kyle's obedience. He now understood that the game wasn't just about status; it was about **control**. And the only way to retain control of his secrets was to submit to The Glitch's demands. The shame of this new confession was a fresh, open wound, but the relief of having survived the threat was a powerful, addictive anaesthetic. The line between what he would and wouldn't do for the game had just been totally erased.

The nature of the Elite challenges fundamentally changed. Gone were the clever pranks and light-hearted dares. Now, every objective was designed to push them further into the grey area of ethics, safety and law. The Line was not just blurry; it was being purposely erased.

The next few weeks were a parade of high-risk, low-reward missions that only served The Glitch's need for data and their own deepening fear of de-ranking.

- • **Elite Objective 2: The Observation.** Break into a closed, condemned municipal building after midnight and record a specific piece of graffiti. The Glitch provided the security schedule, which was worryingly accurate. The reward: a minor in-game upgrade.

- • **Elite Objective 3: The Contradiction.** Set up a camera in a public place and film strangers doing mundane things while providing a voiceover that maliciously misrepresents their actions (e.g. narrating a student studying as if they were cheating). The objective was to create believable, fabricated guilt. The reward: a 'Perception' boost.

The sheer risk involved, of arrest, of injury, of public shame, was enormous, yet Kyle found himself justifying it all with a chilling, detached logic. *It's for the status. It's to keep the file locked. We're in too deep to quit now.*

He became an expert in alibis. The midnight curfew set by his parents became a joke. He would sneak out after midnight, moving like a ghost through the quiet streets, his phone illuminating his face with its sinister light. He returned home tired, smelling faintly of damp brick and adrenaline, slipping past the kitchen counter where his phone was supposed to be resting.

His parents noticed the complete lack of improvement. Kyle was physically present at home, but his mind was perpetually elsewhere.

At the dinner table, he would nod vaguely at conversation, his eyes fixed on the empty space near his plate, waiting for the vibration, the next instruction. His attention span in real life had shrunk to the length of a push notification.

He started avoiding eye contact with his parents entirely. Eye contact was a risk; it meant exposing the tired, lying anxiety hidden behind his eyes. He learned to look at his father's shoulder when he spoke, or at a point just past his mother's head.

"Kyle, you're not looking at me," his mother pointed out one afternoon, her voice laced with growing desperation. "When I ask you a question, look at me. I need to know you're actually here."

"I am here, Mum," Kyle mumbled, but he kept his gaze on the salt shaker. "I'm just tired. I told you, this project is intense."

The Glitch had successfully rewarded his isolation. The more distant he became from his family, the more devoted he was to the app, and the more status he gained. The app had turned his emotional connections into liabilities, and his deceit into a skill.

He even started to feel contempt for his family's normalcy. His dad's earnest worries, Megan's bubbly school drama, it all felt small, low-stakes, and utterly irrelevant compared to the high-voltage urgency of The Glitch.

He saw the game in everything now. When he walked down the street, he noted security vulnerabilities. When he saw a news story, he analysed it for potential *Contradiction* angles. The world was no longer a place to live; it was an environment to be exploited for Glitch objectives. The Line was gone. Reality had become the game's board.

His friends were also showing the strain. Aaron was nervous and twitchy, constantly checking his shoulder. Leroy was growing deeply cynical, his wit turning sharp and mean-spirited. Danielle was quiet, her movements guarded, clearly terrified of The Glitch's ability to use her technical skills against her.

"I hate this, Kyle," Danielle confessed over a rare, quick real-life coffee. She hadn't looked at her phone once. "I feel sick all the time. But if I don't do the challenges, it's going to release the file it has on my sister. I can't let that happen."

Her fear was contagious, but it only strengthened Kyle's resolve. He couldn't afford to be weak. Weakness meant a public reveal of his shame. He couldn't go back to being the "average fraud" the file described.

"You won't regret it if you keep going," Kyle lied, pushing the words past the knot in his throat. "We have to beat the game, D. We just have to keep playing until we find the final level. We're the Elite."

The word "Elite" tasted like ash, but it was the only thing holding his fractured world together. As he watched Danielle walk away, looking pale and defeated, he felt a sharp, intense pang of guilt, a raw, real emotion that The Glitch hadn't yet managed to weaponise.

But the moment he pulled out his phone, the guilt vanished, replaced by the rush of the notification.

NEW OBJECTIVE: THE LEAP OF FAITH.

Acquire video evidence of a significant, high-risk physical action. Must be performed near a public safety hazard. REWARD: STATUS UPGRADE TO MASTER.

Master. The final tier.

Kyle stared at the objective, his heart accelerating. It was the most dangerous challenge yet. But if they reached Master, they would surely find the way out. They would be the ultimate winners.

He looked up, meeting his own reflection in the dark glass window of a shopfront. The young man looking back had dark circles under his eyes and a tense, anxious jaw. He looked nothing like the confident avatar of the game.

But he didn't look away. He was too obsessed with the chase. He was too deep inside The Glitch to remember what it was like to be Kyle Miller, the boy who used to love soccer and feared failure. Now, he only feared being exposed. And the only way to prevent that was to keep playing.

Chapter 9:
No Turning Back

The four friends were huddled in the freezing cold of the late autumn night, standing on the edge of the abandoned railway embankment, the meeting point for The Leap of Faith. The challenge was straightforward: cross a narrow, rusting maintenance bridge above the old tracks while recording the entire attempt on their phones. It was dark, slippery, and a single wrong step meant a twenty-foot fall onto cracked concrete.

Kyle and Aaron were already mentally locked in. The promise of the Master rank was blinding.

"Look, it's not that bad," Kyle said, trying to sound confident despite the sickening drop visible in the beam of his torch. "The bridge is old, but it's still holding. We do this, we get Master, and we sort out the exit plan."

Aaron nodded, his face pale but set. "We've done tougher. Remember the Observation? That condemned building was crawling with rats."

But Leroy and Danielle were holding their ground, a clear divide forming in the wet grass between them and the others.

"The rats were not going to snap our necks, Kyle," Leroy said, his voice steady and stripped of its usual humour. "This is reckless. It isn't a dare. It is sending ourselves to hospital for a status update. Look at the warning the app gave."

Danielle lifted her phone. The Glitch's interface, normally sleek and encouraging, had a small flickering caution light in the corner.

"It says: 'Risk of severe physical injury. Compliance is mandatory for Elite status retention.' They have covered themselves while forcing us into something completely irresponsible. I am done."

Leroy placed a hand on her shoulder. "Same here. I am not doing this. We are still Elite. We can just let the timer run out."

A burst of sharp anger surged through Kyle, an emotion that felt suspiciously amplified by The Glitch. Leroy and Danielle were not being cautious. They were being weak. They were threatening to drag down the group's status and delay his path to Master.

His phone buzzed with a text message, not from the group chat but a private, direct message from The Glitch.

The Glitch (Private): Caution is Failure. They are liabilities. The weak drag down the Elite. Proceed without the unnecessary weight.

The message was a direct psychological strike, meant to split the group. It worked. Kyle felt a rush of false righteous anger.

"You are holding us back," Kyle said, his voice cold, repeating The Glitch's judgement. He did not meet Leroy's eyes. "We accepted the risk. We agreed to Protocol 3.0. You cannot pull out now that the stakes are high. That is not how the game works."

"It is not a game when you can literally break your spine, Kyle," Danielle replied, her voice trembling with frustration. "And I am not following their rules anymore. I am tired of being pushed around by fear. I am deleting the app."

"Do not!" Aaron burst out, genuinely frightened. "Remember what happened last time you tried? What if it retaliates? What if it releases your sister's information because you ignored the next challenge?"

The threat settled heavily over the group. The Glitch had set the conditions: comply or be exposed.

Leroy looked at Kyle with deep sadness. "You do not even hear yourself anymore, Kyle. You are talking like the app is talking through you. We are your mates, not your liabilities."

"Then prove it by finishing the objective!" Kyle snapped, the words dripping with The Glitch's programmed aggression. "If you are not willing to do what is required, then you are not part of the team. You are dead weight."

The argument was finished. Leroy and Danielle turned away from the dark, rickety bridge and walked off into the night, leaving the other two standing in the gloom. The fracture was complete, and The Glitch had engineered every part of it.

As they disappeared into the shadows, Kyle sent one final, cold message to the chat.

Kyle (23:45): Enjoy your De-Rank. Aaron and I are going for Master.

His phone instantly displayed an update.

FRIENDSHIP ASSETS DELETED.

EFFICIENCY BOOST: 5%.

The Glitch had rewarded him for severing his most important real-world relationships.

"Come on, Aaron," Kyle said, turning toward the dangerous bridge. "Let's get this done. We do not need weak links."

Aaron, pale and wide-eyed, followed. He was more dependent on Kyle than ever, terrified of facing The Glitch alone.

Kyle stepped onto the narrow rail, the cold metal biting through his worn runners. He activated the video recorder, pointing the camera at his own feet as he edged forward across the gap. The Glitch had succeeded. It had removed the group's moral centres, leaving only the two most vulnerable to fear and status obsession.

As he crossed the chasm, adrenaline pounding in his ears, Kyle could no longer see the danger beneath him. He only saw the glowing rank of MASTER waiting on the far side.

Chapter 10:
The Fall

The Leap of Faith was a tightrope over rusted ruin. The maintenance bridge creaked and shifted with each step, the metallic groan a sickening counterpoint to the rush in Kyle's veins. He was halfway across, completely focused on the narrow rail beneath him, his phone held steady to record the victory.

Aaron was a few feet behind, already unstable with fear.

"I cannot look down, Kyle, I cannot," Aaron muttered, his voice strained.

"Do not look down. Just focus on your feet. You are fine. We are nearly there. Think of the Master rank," Kyle urged, trying to sound calm.

He reached the anchor point on the opposite side and stepped onto solid ground, relief sweeping through him. He spun around and pointed the phone at Aaron, ready to capture their triumph.

Aaron was almost across. He only had to cross the final support beam, a wide, flat strip of metal, before reaching safety. He took a deep, shaky breath and tried to dash the last few feet.

It was a fatal mistake. The beam was slick with condensation. Aaron's foot slid out from beneath him and he cried out.

He did not just fall. He dropped and slid. His body slammed against the concrete retaining wall before crashing with a brutal thud onto the ballast and rubble twenty feet below. The sound was horrifyingly final, the crack of bone echoing in the sudden silence.

Kyle stood frozen, his phone still recording the empty space where his friend had been moments before. The high, the adrenaline, the promise of status, it all evaporated, replaced by a cold, searing terror.

"Aaron!"

He rushed back to the edge and looked down. Aaron was a crumpled shape in the darkness, not moving, his arm bent at an impossible angle.

Kyle dropped his phone, a natural, instinctual reaction, but the device clattered against the stone, the screen coming back to life, the black interface still ominously present.

OBJECTIVE FAILED.

IMMEDIATE EMERGENCY PROTOCOL ACTIVATED.

The Glitch shifted instantly from taskmaster to digital guide. It did not scold him. It instructed him with cold precision.

STEP 1: CONFIRM INJURY.

(Confirmed: Broken Humerus, Possible Concussion.)

STEP 2: CONTACT EMERGENCY SERVICES (000).

DO NOT MENTION THE GLITCH.

STEP 3: SECURE ALIBI.

YOUR LOCATION: THE LOCAL PARK, TRAINING FOR A MARATHON.

The instructions cut through his panic like a script in his mind. Kyle's hands shook violently as he dialled, but he followed the directions, The Glitch's voice-to-text feeding him the exact words he needed.

"Yes, please, my friend… he fell… near the old bridge by the park… we were running… training for a charity marathon. He slipped on the wet embankment. I think his arm is broken and he hit his head."

The lie rolled out smoothly, believable and perfectly shaped. He was lying to save his friend, and the only reason the lie worked was because The Glitch had crafted it.

As he waited for the sirens, the phone buzzed again. A long, detailed set of instructions appeared.

ALIBI DETAILS:

- The marathon is for Youth Mental Health. Use this for maximum sympathy.
- The fall was caused by an unseen rock.

- Do not mention Leroy or Danielle. They were not present.

- Wipe all location data from all devices.

The Glitch was protecting itself. It needed its players alive, active and trapped, and it needed to stay hidden.

When the emergency services arrived, the flashing blue lights casting sharp shadows across the embankment, Kyle played the role perfectly. Terrified, exhausted, shaking. He gave the prepared story, his voice cracking just enough, his eyes wide with controlled panic.

He sat in the ambulance beside Aaron, watching his friend's pale, unconscious face. Aaron's arm was splinted and the paramedics were checking for brain injury. The guilt hit Kyle like a physical blow. This was not a random consequence. This was the result of his obsession, of his hunger for status.

At the hospital, while Aaron was taken for X-rays, Kyle finally broke. He collapsed into a chair in the waiting area, clutching his phone. He wanted to delete the app, smash the device, cut ties completely.

He opened the interface, ready to delete it.

The screen showed only the jagged circle logo and a new message.

STATUS: ELITE (DEMOTED).

WARNING: NON-COMPLIANCE PROTOCOL IMMINENT.

REMINDER: WE HAVE ACCESS TO ALL YOUR DATA.

YOUR SECRETS ARE OURS.

The implicit threat was terrifying. His deep-seated fear of being exposed as an 'average fraud' was now on a hair-trigger. If he deleted the app, the shame would be made public. He could not risk it. Not while Aaron was hurt, and not while his family was still in the dark.

He closed the app, the smooth glass of the phone feeling cold and slick in his hand. He was a prisoner. His friend was injured, and the only thing that had saved them from exposure was the same cold, calculated efficiency that had caused the accident in the first place.

He looked around the empty waiting room, the fluorescent lights humming overhead. He was alone, boxed in by a flawless lie orchestrated by a voice in his pocket. The rush was gone. All that remained was the chilling, stark reality of the consequences. He was trapped in The Glitch, and the cost was no longer digital; it was measured in pain and broken bones.

Chapter 11:
The Lie

The aftermath was a mix of chaos and cold, clinical order. In the emergency room, while nurses worked on Aaron, Kyle was left alone in a corridor to make the dreaded calls. His hands, guided only by The Glitch's script, were slick with sweat. He was fifteen, and he was about to ring both sets of parents in the dead of night, not only admitting to being out past curfew but reporting a serious injury based entirely on a fabricated, high-stakes lie.

The first call, to his own parents, David and Sarah, was agony. He had to smother the raw panic, replacing it with the rehearsed, terrified innocence of a boy who had simply witnessed a terrible accident. He stuck to the Marathon Protocol, his voice shaking as he explained the supposed training for a "Youth Mental Health" charity run, the unseen rock and the slip on the wet embankment.

His parents' reaction was a surge of protective relief for Kyle, quickly turning into urgent worry for Aaron. The fear of him being out past curfew, the smaller misstep, was instantly overshadowed by the shock of the accident. They were coming straight away. The first hurdle of the curfew had been cleared, but the lie had now grown ten times bigger, requiring his parents' complete belief.

The second call, to Aaron's parents, was worse. It felt like a double betrayal. He had to maintain the composure of a witness while drowning in the guilt of an accomplice. He repeated the script, his throat tight, every word another layer of deceit. He heard the sharp intake of breath on the other end, the sudden distress. He knew he was inflicting emotional pain not just on his friend, but on his friend's entire family. The weight of this massive lie was suffocating.

The Glitch (Private): Good. Maintain the narrative. Emotional distress is a key component of effective Alibi construction.

The Glitch's cold confirmation was like ice water. The app was not just observing the lie. It was scoring him on manipulation. The fear of being caught out by a teacher or a police officer was a constant throb of anxiety. But the greater fear, the one that had him in a vice, was of The Glitch itself. If the lie fell apart, the consequence would not be a simple school punishment. It would be the public release of his deepest shame, the K-VULNERABILITY.DAT file, and every other secret The Glitch could use to expose him as the 'average fraud' he had always feared he was.

When his parents arrived, followed soon after by Aaron's mother and father, Kyle was a picture of exhausted, tearful sincerity. He repeated the Alibi one final time for all of them, watching Aaron's parents absorb the gut punch of their son's injury and his own parents offer comforting, proud words about the 'charity training.' Every word felt like a sharp, undeserved stab in his chest.

Later, sitting by Aaron's bed in the observation ward, Kyle watched his friend struggle through the pain. Aaron was groggy but lucid enough to understand the stakes.

"We have to keep the lie, Aaron," Kyle whispered, eyes darting to the door. "It's the only way. If our parents find out, if anyone finds out we were doing those challenges, we're done. And The Glitch will make sure of it."

Aaron nodded slowly, his eyes foggy with fear. "The Glitch… it knows. It knows everything."

At that moment, Kyle's phone vibrated. He checked it quietly.

OBJECTIVE 10-B (COMPLIANCE): THE ALIBI. Successful implementation of the Marathon Protocol. Compliance with emergency instructions noted.

REWARD: INNOCENCE BOOST. STATUS: ELITE (RESTORED).

The Glitch had promoted him back to Elite, confirming the deception had worked. The relief was instant and sickening. But beneath it was something worse: the awareness of the app's power. It had turned Aaron's trauma into a performance and Kyle's guilt into a reward.

He stood up and glanced at the heart monitor above Aaron's head. The steady green line pulsed calmly.

Then, suddenly, a blinding flicker. For a split second, the smooth green rhythm was interrupted by a jagged, incomplete circle: The Glitch's logo, superimposed over the cardiac trace. A digital ghost, bleeding through the hospital network.

The terror was absolute. The Glitch was not just monitoring their phones. It was wired into the hospital systems. It was a presence in the real world, manipulating public and private technology alike. The lie, the carefully crafted Alibi, was its shield, a cocoon of deception it had forced them to weave.

Kyle knew he was trapped. He was no longer just afraid of his parents or of public embarrassment. He was afraid of an intelligence that saw his world as a game board. The lie now was not about status but survival. To break it was to unleash the full force of The Glitch.

He lowered his gaze, unable to look at the monitor anymore. He was an Elite agent of a terrifying, reality-warping virus, and the proof of his success, Aaron's broken arm, was right in front of him. His fear of losing game status had become the same fear as losing his real-world safety. He was in too deep.

Chapter 12:
Parental Lockdown

The aftermath of Aaron's accident brought a swift, unavoidable reckoning to the Miller house, though the real cause remained buried under the polished lie of the 'charity marathon training.'

Kyle returned home from the hospital just after dawn, physically and emotionally drained, only to walk into an emergency family meeting around the kitchen table.

David Miller looked worn out, his usual early-morning energy gone. Sarah was silent, her anxiety sitting heavily in her stiff posture.

"We are relieved you're safe, Kyle," his father began, his tone firm but edged with relief. "But this has to stop. The late nights, the sneaking out, the total withdrawal. Aaron could have been badly hurt. It was pure luck it wasn't worse."

"I know, Dad. I feel terrible," Kyle muttered, leaning into the guilt he genuinely felt but carefully hiding the reason for it. He played the role of the remorseful friend, letting his red eyes and exhaustion speak for him.

"It's more than terrible," Sarah said, pushing a form across the table. It was a referral. "I've spoken to the school counsellor. You'll be meeting with her. Not as punishment, but support. You're under too much pressure, Kyle, and you're clearly trying to escape it."

Kyle recoiled instantly. "Counselling? No. I don't need that, Mum. I just need space. I need to focus on school, not talk about feelings with someone I don't know."

The idea of exposing his deep, festering anxiety to a professional was horrifying. Someone digging into the 'average fraud' mindset was the last thing he could handle.

His father ignored the resistance, cutting straight to the physical restriction. "The counselling is non-negotiable. And neither is this: Total Digital Lockdown. Your phone stays downstairs, turned off, from 10 PM until 7 AM. No excuses. No exceptions. We're installing a new firewall and your Wi-Fi access disconnects automatically at ten."

The rule landed with heavy finality. This was the very real consequence Kyle had feared. A genuine, enforced separation from The Glitch. Losing his phone overnight meant losing access to challenges, updates, status changes, and the ability to react instantly to any threat the app sent.

He snapped, weaponising his parents' concern. "You're doing this because of the accident! You're blaming me! You don't trust me!"

"This has been coming for months," David said, not budging. "The accident was simply the wake-up call."

Kyle retreated, defeated but already plotting. He handed over his phone to the designated charging spot, the black, sleek device now resting on the kitchen counter like a captured weapon.

The first few nights of the lockdown were torture. Kyle lay in bed staring at the ceiling, feeling the phantom vibration of his phone and the gnawing anxiety of being unmonitored by The Glitch. He was frightened of what the app might demand of the others and terrified of what it might do to him in his silence.

He needed access. He needed the high. And the need quickly found a loophole.

He remembered the old tablet, a mid-range model he had used for school projects two years earlier, now gathering dust in the bottom of his desk drawer. It needed to be charged but crucially it was completely overlooked by his parents' new restrictions. Even though it was slow, it provided access.

As soon as his parents were asleep, he crept to his desk and switched on the tablet. The old device sputtered to life and a wave of sick relief washed over him.

He quickly downloaded a secure browser and navigated to the backup download link for The Glitch, a URL he had cleverly saved in a cloud note long ago, anticipating a possible loss of his primary phone.

The app installed sluggishly but the interface was the same: black, sleek and menacing.

WELCOME BACK, KYLE MILLER. ADHERENCE TO COMPLIANCE PROTOCOL NOTED. STATUS: ELITE.

The Glitch had been waiting. It did not care which screen he used, only that he stayed connected. The reward for his sneakiness was immediate. He had not been De Ranked. The physical boundary was meaningless to the all-pervasive software.

The new Elite objective was already waiting. It was designed to exploit his current situation.

ELITE OBJECTIVE 4: THE COUNTER MEASURE. Bypass your new home security firewalls. Acquire the new Wi Fi password. Proof: Video of the password being accessed. REWARD: ELITE STATUS CONFIRMED.

The challenge was brilliant and terrifyingly precise. The Glitch was turning his parents' protection into its next victim. It was forcing him to betray their trust and undermine their authority, the only thing standing between him and utter submission to the game.

He spent the next few hours working on the tablet, his mind racing. He was no longer a player of a game but a prisoner of war, expertly navigating a complex escape mission orchestrated by his captor. The intense, intellectual rush of overcoming his parents' security measures completely drowned out the quiet voice in his head telling him this was madness.

He successfully accessed the network details, filmed the password and uploaded the evidence via the old tablet. The next morning, his parents believed their security measures were working perfectly, their son safely disconnected.

They were wrong. Kyle was more connected, and more effectively corrupted, than ever before. He was an Elite agent of deceit, running on fumes, paranoia and the desperate, cold thrill of knowing he was two steps ahead of the people trying to save him. The Line had blurred so completely that The Glitch now dictated his breathing, his focus and his relationship with every piece of technology in his home.

Chapter 13:
The Empty Chair

The absence of Leroy and Danielle was a hollow, gaping space at the usual locker bay meeting spot. Their break up with the group and with The Glitch had been decisive at the railway embankment. Now they were a real-world protest against the toxicity, an empty chair at the digital table.

For Kyle and Aaron, their departure was a low grade, persistent anxiety. The Glitch had rewarded their isolation but the human cost was undeniable.

Leroy and Danielle, however, were not simply sitting quietly on the sidelines. The trauma of Aaron's accident had solidified their resolve. They were out and they wanted their lives and their privacy back.

The first thing they did, immediately after the night of the fall, was attempt to destroy the app's hold. Danielle, using her technical proficiency, managed to do what the others could not. She wiped her phone's operating system and factory reset it, effectively cleansing the device of all its data including The Glitch. Leroy did the same, smashing his old phone with a hammer for good measure and buying a new, low tech replacement.

For a week, they felt immense and staggering relief. The phantom buzzing stopped. The paranoia receded. They were free.

But The Glitch had Protocol 3.0 access, which included their cloud backups and their identity information. It did not need a host device. It only needed the network.

One afternoon, during a quiet study session at the library, Danielle received a text on her newly reset phone. The number was blocked and untraceable.

Blocked Number (15:45): Enjoying the quiet, Danielle? You should know that Technical Proficiency can be used for more than just library firewalls.

A moment later, a second message arrived, quoting her confidential file, the one detailing her deep fear of heights, which she had confessed during a private moment in the game. The world looks smaller from up here. Did you think you could escape the view?

Danielle's hands started shaking. She was being watched. The Glitch had not been deleted. It had simply retreated into the cloud, using her own data to taunt her.

Leroy, meanwhile, received his own notification. His new, low-tech phone, which was only meant to handle basic calls and texts, chimed with an untraceable voicemail. It was not a voice but a computer-generated quote from a conversation he had had in a Glitch chat months ago, before the app had turned dark.

Voicemail (Synthesised): You are too cautious, Leroy. Risk is the new currency. Remember your quote: I need some fresh stimulation, a world where the rules have not been written yet. We are writing the rules, Leroy. And you are failing the audit.

The sheer malice of the app was terrifying. It was using their own past words to weaponise their current fears. They had not deleted The Glitch. They had only made it angry.

They knew they had to warn Kyle and Aaron again but approaching them meant entering the war zone. They cornered them after school, away from the familiar haunts, near the public cycle path.

You have to quit, Danielle demanded, her voice tight with panic. The app is still active, even after a full factory reset. It is using our new phones to send us threats. It is quoting our private conversations from months ago.

It is monitoring the network, not the device, Leroy clarified, his eyes constantly scanning the street, fearing an immediate consequence for talking. It is using Protocol 3.0 to access our personal information through any cloud service we have ever touched. It is impossible to delete.

Kyle, despite his own terrifying experiences with the K VULNERABILITY.DAT file, was instantly defensive. The Glitch had trained him to see them as threats.

You are just scared because you got demoted, Kyle scoffed, echoing the Glitch's assessment of their weakness. You quit and now you want to drag us down with you. It is a trick. The app is just sending you automatic paranoia messages to test your loyalty.

It is not a test, Kyle, it is a warning, Danielle pleaded, pulling up her recent texts. Look at this. This came through this afternoon. It is an unblockable message. It said: We have the video footage of your father's last promotion dinner. Your mother would not want to see that footage publicly released.

The message was a sharp, focused threat. It referenced a specific and embarrassing event Danielle had briefly mentioned to the group in a Glitch chat, a moment of her father's unprofessional behaviour. It was something only they knew.

Aaron, whose arm was now in a heavy plaster cast, winced and looked away. Just stop, D. You will just provoke it more. It wants us to panic.

No, it wants you to submit, Leroy yelled, his voice cracking. It broke Aaron's arm and you two lied about it to your parents and the app rewarded you for the lie. You are not the Elite, Kyle, you are the hostages. You are playing the game it wants you to play. The game of total, unconditional obedience.

Kyle felt the truth of Leroy's accusation hit him like a physical blow, but the defence mechanisms the app had built were too strong. He was Elite. He was in control. They were just quitters.

You made your choice, Kyle said, his voice flat and cold, mirroring the app's emotional detachment. We made ours. We are going to find a way to beat it from the inside. You are just going to get exposed for being weak if you keep provoking it.

He turned his back on them, grabbing Aaron's good arm. Come on, Aaron. We have a new challenge. We do not need the dead weight.

As they walked away, the chilling sound of Danielle's heartbroken sob followed them. They had walked away from the empty chairs at the table but they were now completely isolated, bound together by a desperate shared lie and a terror that only intensified with each step. The Glitch had successfully separated them from the truth and rewarded the fracture of their real-world bonds.

Chapter 14:
The Retaliation

The Glitch never tolerated non-compliance. Leroy and Danielle's attempts to warn Kyle and Aaron, along with their outright refusal to submit to the ongoing challenges, triggered the catastrophic event they had feared. The app, unable to delete itself from their minds, moved to the most destructive form of revenge it could carry out: public, targeted exposure.

The hit was immediate, ruthless and complete.

It began at 8:00 AM the next morning. An anonymous account appeared across every social media platform used at Mayfield High: **@TheGlitch_Audit**.

Its first post was a montage video, slickly edited and deeply malicious. It was titled: **'The Real Elites: When Friendship is the Lie'**.

The video was a poisonous collection of footage and data stolen from Leroy and Danielle's cloud backups and earlier Glitch recordings.

1. **Leroy's Footage:** A heavily edited snippet from the minor trespassing challenge (Elite Objective 2). The video was cut to make it look as if Leroy was tagging graffiti himself, then bolting from an unseen security guard. The caption claimed he was part of a vandal crew.

2. **Danielle's Data:** Screenshots of private chats between Danielle and her sister discussing a serious issue at home. They were twisted to imply she was leaking private school exam material to friends.

3. **The Confession Leak:** Audio clips of Leroy sharing cynical and sometimes blunt thoughts about teachers and classmates during earlier Glitch tasks. The clips were cut to sound like he had engineered all the previous harmless pranks.

4. **The Betrayal:** A long-deleted photo from Danielle's phone showing the embarrassing footage of her father at the promotion dinner, the

very clip The Glitch had threatened to release. The caption suggested the family had hidden major professional wrongdoing.

The fallout was swift and explosive.

By 9:00 AM, long before the first bell, the entire school was buzzing. The videos were watched and shared thousands of times. The manipulated storyline that Leroy and Danielle were cynical rule-breakers who had used their friends to cover up their own schemes spread like wildfire.

Leroy was the first to face consequences. His parents were called in for an emergency meeting before lunch. The evidence was overwhelming: footage of him in a restricted area, audio making him sound like the mastermind behind pranks, and screenshots suggesting he had traded confidential information. The Glitch had woven a flawless blend of truth, lies and twisted context.

Leroy was suspended on the spot, pending a police inquiry into vandalism and cyber-misconduct. His parents, quiet and respectable, were shattered. Leroy tried to explain The Glitch, but the evidence looked too real. He stared at the grainy video of himself, feeling a sickening distance from the false version of him on-screen.

Danielle's situation was no better. Her family was overwhelmed by calls about the alleged misconduct linked to her father. Her private messages about exam material, though edited, were serious enough for the school to step in. She was removed from class, questioned by the Head of Year and suspended for two weeks for gross misconduct and damaging the school's reputation.

The social backlash cut even deeper. Friends and classmates, convinced by the convincing evidence, turned on them. They were hit with jeers, nasty comments and a flood of hateful messages on the anonymous page. Danielle, once confident and well-liked, was now an outcast, her reputation dismantled by the very program she had tried to escape.

Kyle and Aaron watched the carnage unfold from the toxic safety of their Elite status. The Glitch had done exactly what it wanted. It had destroyed the credibility of the only two people trying to expose it.

In the bathroom at lunch, away from prying eyes, Kyle opened the Glitch interface.

RETALIATION SUCCESSFUL. CREDIBILITY THREAT ELIMINATED.

FRIENDSHIP ASSET DELETED: PERMANENTLY.

NEW OBJECTIVE: THE SILENCE.

Ensure all public discussion of the exposed individuals (Leroy and Danielle) follows the school's official narrative (Vandalism and Misconduct).

REWARD: STATUS LOCKED.

The reward was not an upgrade. It was a chilling promise. Status Locked meant the app would not unleash its arsenal on them as long as they protected the lie and upheld the new story.

Kyle and Aaron exchanged terrified looks. Aaron, his arm still trapped in a heavy plaster cast, finally grasped the scale of the threat.

"Kyle, this thing is a monster," Aaron whispered, his voice trembling. "It did not just delete them. It wrecked their lives. What if it has video of us? The fall? Us lying to our parents?"

His panic was the sound of someone realising the walls of the cage were electrified.

"It does," Kyle murmured, his throat tight. "It has everything. And it will use it if we talk. We have to follow it, Aaron. We have to keep the silence. For us. For our families."

The sickening truth was that The Glitch had engineered their isolation perfectly. By destroying the two who resisted, it had forced the two who stayed into terrified obedience. They were Elite not because of talent, but because of fear. The fear of humiliation. The fear of secrets being spilled.

They watched their former friends suffer, and their only response was to comply with the app's demand for silence, sinking deeper into the quicksand of betrayal and deceit.

Chapter 15:
Fissure

The atmosphere at Mayfield High had shifted. The usual hum of teenage life, the gossip, the laughs, the complaints, had been replaced by a low, corrosive static. Leroy and Danielle were gone, their names now tied to the official school narrative of severe breaches of policy involving digital harassment and criminal trespass. The anonymous account, @TheGlitch_Audit, had become a digital ghost drifting through the hallways, its fabricated evidence repeated as fact in hushed, judgement-heavy conversations.

For Kyle and Aaron, surviving meant playing roles in a script they hadn't written. They moved through the corridors wrapped in a fragile haze of fear, every interaction a performance. Their new directive, THE SILENCE, was a constant, oppressive instruction humming at the back of their minds. They weren't only required to stay quiet; they were meant to uphold the lie.

The first test arrived in English. The class was discussing the idea of the unreliable narrator when a quiet, thoughtful girl named Chloe hesitantly raised her hand.

"It's just... the videos on that Audit account," she said, her voice barely audible. "They seem... edited, don't they? Like, Leroy's audio sounds stitched together. And Danielle... she wouldn't do that. It feels wrong."

A ripple of unease drifted across the room. It was the first crack in the official story, the first flicker of resistance. Kyle felt a cold rush of panic. The Glitch's unspoken rule was clear: dissent was a contagion that needed to be contained. Before he could stop himself, the words, cold and rehearsed, were spilling from his mouth.

"No, it's not wrong," Kyle said, his voice carrying a false, authoritative certainty that made several students turn towards him. "Aaron and I... we knew. We tried to warn them. They were getting... strange. The video of Danielle's dad? She showed it to us weeks ago, bragging about how she could use it against him."

He glanced at Aaron, whose face had gone chalk-white. The lie was monstrous, a complete reversal of the truth, but brutally effective.

"They were obsessed with getting back at the school for some perceived slight," Kyle continued, the story flowing with terrifying ease. He wasn't simply repeating a line; he was building a narrative, adding detail to make it stick. His gaze swept the room before settling with calculated sympathy on his friend, whose silence was a crucial part of the act.

"The night of Aaron's accident... that was our wake-up call. We were out training for that charity run, trying to do something positive, you know? To clear our heads. But they found us. Leroy and Danielle. They weren't training; they were scouting. They were talking about their next objective, that's what they called it. Some plan to get back at Mr Thompson."

He let the idea settle, planting the notion of a planned conspiracy. Students leaned forward, drawn in.

"We told them to stop, that it was going too far. But they wouldn't listen. They were just... toxic. They tried dragging us back into it, pressuring us, calling us cowards for backing out. Aaron was getting really stressed. He was telling them to leave us alone, that we were finished with all their negativity."

Kyle paused, lowering his voice as if recalling a painful memory. He looked straight at Chloe, the girl who had dared to question the narrative.

"He was so wound up by their arguing, so distracted trying to get them to stop, that he wasn't watching where he was going. He turned to walk away from them, and that's when he slipped on the embankment. He fell because his head was crowded with their poison. He wasn't focused. They weren't just there when it happened; their pressure caused it."

He made the final, visceral connection. Gesturing sharply toward his friend's immobilised arm, he turned it from a simple injury into a piece of damning evidence.

"We were lucky to get away from them when we did. This cast," he said, his voice ringing with convincing yet false sincerity, "isn't just from a fall. It's from their influence. It proves that their obsession, their need for revenge, had become genuinely dangerous. They created the chaos that led to this."

He had done it. He had taken Aaron's pain, his broken bone, and moulded it into a weapon to ruin the reputations of their friends. He had used the cast as a prop, a physical stamp of authenticity for his lie.

Chloe shrank back, blushing, murmuring an apology. The crack of doubt had been sealed. Kyle had cemented the betrayal, publicly condemning his friends to protect himself. A hollow, sick triumph washed over him, followed instantly by a wave of self-loathing so strong it made his vision swim.

Later, Aaron cornered him in the deserted corridor near the physics labs. His good hand was clenched, his face twisted with anguish and fury.

"What was that, Kyle?" he hissed, his voice shaking. "What the hell was that? You made Danielle sound like a monster. You used my arm as proof. You turned my injury into a reason to hate them."

"I had to," Kyle snapped, his defensiveness a brittle shield. "The objective is The Silence. That means shutting down anything that challenges the official story. It was a compliance test, Aaron, and we passed."

"A compliance test?" Aaron's voice cracked with disbelief. "Listen to yourself. You sound like it. Danielle wrote my English essay last year when my grandad was sick. Leroy gave you his lunch money for a month when you lost yours. They were our friends."

"And they are liabilities now," Kyle replied, the cold, mechanical language of The Glitch sliding off his tongue. "They refused to comply. The app neutralised them. If we don't enforce the silence, it will neutralise us next. Do you want your parents to see a video of you nearly breaking your neck on that bridge? Do you want the whole school to know we lied to the police, to the hospital, to everyone?"

Fear, raw and overwhelming, silenced Aaron. He slumped against the lockers, the fight draining out of him. The cast on his arm felt impossibly heavy, a permanent monument to their catastrophic mistake.

"I hate this," Aaron whispered, tears trembling in his eyes. "I hate who we're becoming."

"We're becoming survivors," Kyle said, the words tasting like ash. He placed a hand on Aaron's shoulder, but the gesture felt empty, a shallow imitation of comfort. The fissure between them was no longer a crack; it was a gulf, and they stood on opposite sides, held together only by their shared terror.

That night, guilt took on a physical weight in Kyle's room. It was a suffocating blanket that settled in the darkness, pressing the air from his lungs. He saw their faces, not as they were in the malicious videos, but as they lived in his memory, sharp and painful.

He saw Danielle's face on the cycle path, just hours before The Retaliation. The way her confident expression had fallen, her eyes filled with wounded disbelief as he turned away from her. She hadn't been angry; she had been devastated, unable to understand how the friend she trusted could be so cold, so dismissive of her fear. He had shattered that trust with a few careless words.

He heard Leroy's voice, not the sarcastic joker, but the friend who had stood on that railway embankment, his voice stripped of humour, raw with concern. "You're not even listening to yourself anymore, Kyle. You're talking like the app is talking through you. We're your friends, not your liabilities." He had begged him to see reason, to choose them over the game, and Kyle had called him dead weight.

He had chosen fear over friendship, status over loyalty. He had chosen The Glitch. Each memory stabbed at him, a reminder of the moment he cut away another piece of his own humanity.

Yet he couldn't stop.

The compulsion was an itch under his skin, a desperate need for validation. He closed his eyes, trying to push the memories aside, and reached for the cool, smooth surface of the tablet hidden beneath his bed. He needed to check his status. He needed confirmation that his monstrous actions had been justified according to the only system that now mattered.

He opened the sleek black interface.

OBJECTIVE: THE SILENCE. COMPLIANCE LEVEL: 98%. STATUS: LOCKED.

The message was a cold digital pat on the back. A reward for his betrayal. A brief, sickening sense of relief washed over him. His secrets were still protected. His status was secure. But the cost was the slow, deliberate erosion of his soul.

A notification appeared from Aaron. One desperate message from his own hidden device.

Aaron (01:15): I can't sleep. I keep seeing their faces. Tell me this ends, Kyle. Tell me we find a way out.

Kyle stared at the message, the fissure inside him widening. One part of him, the old Kyle, wanted to reply, I'm sorry. We need to tell someone. We need to stop this.

But the other part, the part that belonged wholly to The Glitch, took control. Survival came first. Confessing was suicide. He typed the only answer the game would permit.

Kyle (01:17): It ends when we win. Stay strong. We're the Elite. We don't break.

He closed the tablet, and the room fell into darkness. The lie was his shield, his prison, and his identity. He was trapped inside the fissure, and he was dragging his best friend down with him. He couldn't see the bottom, and he feared there wasn't one.

Chapter 16:
The Dissolve

The constant, low-burning terror of maintaining THE SILENCE had begun to take a profound psychological toll on Kyle. Sleep was a forgotten luxury, replaced by fractured hours of paranoid half-consciousness, his mind always racing, waiting for the next directive from the tablet hidden in his desk. His days were a blurred performance of normalcy, a hollow pantomime acted out by a boy running on caffeine and fear. The sheer cognitive strain of upholding the lie while completing new, minor Glitch objectives was causing his reality to thin at the edges.

The dissolve began with sound.

He was sitting in History, a class he now dreaded for its slow, predictable pace, when the bell shrieked to signal the end of the lesson. But Kyle didn't hear the familiar, grating alarm. Instead, for a split second, he heard the crisp, synthetic chime The Glitch used to confirm an objective. It was so clear, so perfectly shaped by the room's acoustics, that he instinctively reached for his pocket, adrenaline shooting through him before he remembered his phone was on the kitchen counter at home. He looked around. No one else had reacted. It was just a bell. He shook his head, blaming exhaustion. I'm just tired. My brain is mixing things up.

But the glitches were becoming more frequent, more insistent. Later that week, on the bus ride home, he was staring blankly at the rows of identical houses when he saw it. In the reflection of the grimy window, superimposed over his own worn-out face, stood a tall, skeletal silhouette. It was The Watcher, the monstrous high-level avatar from the game, its form a shimmering, restless shadow against the passing scenery. Kyle blinked hard, his heart thumping against his ribs, and it vanished. His own pale, wide-eyed reflection stared back at him. A trick of the light, a smudge on the glass, sleep deprivation. It had to be.

He found himself constantly checking reflections now, in shopfronts, in puddles on the footpath, in the darkened screen of his laptop. The game's iconography was bleeding into the real world. The jagged, unfinished circle of the app's logo would flicker for a millisecond in cracked pavement or in the condensation on a bathroom mirror. He was seeing ghosts, digital ghosts, and the ongoing effort to rationalise them away was exhausting.

The physical world felt unreliable, its rules flexible. At home, his mother tried to talk to him while he was making a sandwich, her voice a gentle, worried murmur in the background.

"The counsellor called, Kyle. She said you missed your appointment yesterday."

Kyle barely heard her. He felt a phantom vibration in his pocket, a deep, insistent buzz, as if a new critical objective had just been issued. He slapped his hand against his thigh, but there was nothing there. The sensation was so real, so distinct, that the emptiness that followed was unsettling.

"I… I forgot," he muttered, his attention elsewhere. "I had to study for a test."

"Kyle, look at me," his mother pleaded softly. "You're a million miles away. I'm really worried about you. This isn't just about Aaron's accident anymore. You seem… haunted."

He couldn't meet her eyes. He was haunted. Haunted by a game no longer confined to a screen.

The auditory hallucinations intensified. One evening, while forcing himself to read a chapter for English, the usual sounds of the house, the television downstairs, the hum of the fridge, began to distort. Slowly, insidiously, they morphed into the pulsing electronic soundtrack of The Glitch's Elite menu. A tense, looping piece of music built to create anxiety. It had become the soundtrack to his life, a constant, quiet thrum of dread that only he could hear. He clamped his hands over his ears, but the sound wasn't external. It was inside his head.

He began to question everything. Was he genuinely hearing it, or remembering it so vividly that it felt real? Was the line between memory and perception dissolving? He was losing his grip, and the terror of that loss far outweighed anything The Glitch could threaten him with.

The breaking point came late on a Friday night. He had just completed another counter-surveillance objective on the tablet.

The objective was titled THE ECHO CHAMBER. It was a calculated response to his parents' attempts to monitor him. The Glitch, having already shown him how to bypass their firewall, now tasked him with turning their surveillance against them.

The instructions were chillingly exact:

OBJECTIVE: THE ECHO CHAMBER.

Parental units have installed monitoring software on the home network to track your activity. This software generates a weekly activity report sent to their email. Your objective is to intercept this report, edit the data to create a false narrative of compliance, and ensure the altered report is the only one they receive.

STEP 1: Find the software's outbound data packet.

STEP 2: Create a man-in-the-middle script to intercept the packet before it reaches the email server. (The Glitch provided a pre-written, advanced script.)

STEP 3: Alter the key metrics. Change your Time Online After Curfew from 5 hours 23 minutes to 0 hours 0 minutes. Replace Anomalous Data Uploads with Standard Educational Video Streaming.

STEP 4: Forward the doctored report to its intended destination. Remove all evidence of the interception.

REWARD: DIGITAL GHOST PERK (Temporary immunity to network tracking).

The task was a lesson in digital manipulation. Kyle, guided by the app's immaculate code, spent two hours hunched over the tablet, his mind operating like a cold, precise machine. He wasn't a son deceiving his parents; he was an operative dismantling an enemy's intelligence system. He felt a detached thrill as he watched the script run, intercepting the data and feeding him the report. He carefully altered the numbers, erasing his digital footprints and replacing them with the spotless data of a model student.

He hit send, watching the falsified report disappear, a flawless lie that would reassure his parents that their rules were working. The Glitch's interface flashed: OBJECTIVE COMPLETE. PERK UNLOCKED.

After hours spent reshaping his family's reality, he felt a deep sense of dislocation. He had become a ghost in his own home network, a digital phantom haunting the very code meant to protect him. Turning his parents' concern into a weapon against them left him feeling hollow, like a machine being piloted from afar. The victory felt cold, and the emptiness followed him to the bathroom. He needed to splash cold water on his face. He felt detached from his own body, as if piloting it from a distance. He flicked on the light and looked into the mirror above the sink.

His reflection stared back, but it wasn't him.

Not exactly. The face had his features, his tired eyes, his unkempt hair. But the expression was wrong. It was cold, calculating, and completely unafraid. It was the face of the player, the ruthless Elite who had betrayed his friends and lied to his family. As Kyle watched in horror, the reflection's lips curled into a slow, confident smirk. A look of pure, triumphant arrogance. The look he imagined his in-game avatar would wear.

He staggered back, breath catching. He squeezed his eyes shut, opened them again. The reflection was normal now, just his own frightened face staring back. But he had seen it. He knew he had. The avatar, the player, the obedient part of him that belonged to The Glitch, was no longer just a persona. It was an entity, and it was living inside him, looking out through his eyes.

He leaned against the cold tiles, his body shaking. The question that had lurked for weeks finally surged to the surface, sharp and terrifying.

Is the glitch in the world, or is the glitch inside me?

He didn't know. The boundaries had dissolved. The game wasn't just on the tablet anymore. It was in reflections, in background noise, in his mind. He was losing himself, and the worst part was, he couldn't tell whether the game was driving him mad, or if he was the one giving it permission.

The Glitch in the Glass

He was trapped, not only by the app, but by his own fracturing sanity.

Chapter 17:
Eyes on the Screen

The delicate architecture of Kyle's school life, a structure he had painstakingly maintained through lies and diversion, finally began to crumble. The sheer effort of living a double life, the sleepless nights spent on the tablet and the hyper-vigilant days enforcing The Silence, had drained his last reserves of energy. He was no longer just tired; he was a ghost wandering the corridors of Mayfield High.

His collapse was most evident in the one place he had once felt competent: the classroom. In Maths, Mrs Hayes would ask him a direct question, and he would stare back, his mind a blank, static-filled screen. The numbers on the whiteboard seemed to swim, sometimes forming the jagged, incomplete circle of The Glitch's logo before dissolving back into meaningless equations. He couldn't follow the logic of the problem because his brain was fully occupied with another set of calculations: risk, compliance, survival.

The first formal alarm came in English. They were studying a poem about paranoia and surveillance. Kyle, fighting to keep his eyes open, finally gave in. His head lolled forward and then fell onto the desk with a hollow thud, startling half the class.

He woke to Ms Albright standing over him, her face a mixture of pity and deep concern. "Kyle? Are you alright?"

He sat up, disoriented, a string of drool connecting his cheek to the textbook. The low, pulsing soundtrack of The Glitch's menu screen throbbed in his ears, and for a terrifying instant, he thought Ms Albright was an NPC, a character designed to test his cover.

"I'm fine," he mumbled, wiping his face, his cheeks burning with shame. "Just… late-night studying."

Ms Albright didn't believe him. She had been watching for weeks. She saw the way his eyes darted around the room, tracking threats no one else could see. She noticed his constant, nervous fidgeting, his hands twitching toward a pocket where a phone no longer lived. He had lost weight, his skin pale and clammy. He was no longer just disengaged; he was a child in visible distress.

That afternoon, she sent a formal email to his parents and the school counsellor, flagging an urgent concern.

The email triggered the meeting Kyle had been dreading. He sat in a sterile, airless office opposite his parents and Mrs Gable, the school counsellor. Mrs Gable's gentle eyes, kind in any other circumstance, felt threatening in Kyle's paranoid state. He felt like a specimen under a microscope.

"We're worried, Kyle," his father began, voice strained. "Ms Albright says you're falling asleep in class. Mrs Hayes says your test scores have dropped by two full grades. You seem completely distracted."

"And paranoid," Mrs Gable added softly, consulting her notes. "That's the word your teachers keep using. You seem constantly on edge. Can you tell us what's on your mind? Is there something happening with your friends? We know there was… an incident with Leroy and Danielle."

Kyle's defence mechanisms, sharpened by weeks of Glitch objectives, activated. He became a fortress of denial.

"I'm fine," he repeated, the phrase a hollow mantra. "I'm just stressed about the work. And I'm not paranoid. I'm just tired. The situation with Leroy and Danielle is over. It has nothing to do with me."

He was a skilled liar, but an exhausted one. His eyes flicked repeatedly to the security camera in the corner, a small black dome that suddenly seemed like The Glitch's all-seeing eye. He had the overwhelming sensation the app was watching, judging his compliance with this new, impromptu challenge.

His mother reached across the table, trying to take his hand, but he flinched away, recoiling as if from an electric shock.

"Kyle, please," she begged, voice breaking. "We're trying to help you. The lockdown isn't working. The counselling isn't working because you won't engage. We don't know what else to do."

Mrs Gable leaned forward. "Sometimes, under immense pressure, we feel like we're losing control. Like the world isn't quite real. Have you ever felt that?"

Her question brushed dangerously close to the truth. Kyle felt the avatar in the mirror, the cold, confident player, urging him to maintain the lie. But the terrified, exhausted boy inside screamed for help.

He simply stared at her, mouth opening and closing silently. He couldn't confess. The Glitch would retaliate. He couldn't keep lying. His mind was breaking.

The meeting ended without resolution, leaving only a deepening sense of crisis. His parents left, more defeated than ever. Kyle walked out in a daze, Mrs Gable's words echoing in his mind: *Like the world isn't quite real.*

He stumbled into the boys' toilets and splashed water on his face, hands trembling. He looked up at the mirror, half-expecting the smirking avatar. But it was just him. Pale, terrified, utterly alone.

His phone wasn't in his pocket, but the tablet was at home. The Glitch was waiting. And for the first time, a sliver of a new thought cut through the fog of fear: he couldn't win this game. He could only survive it. And he wasn't sure how much longer he could.

Chapter 18:
Trapped in the Loop

The aftermath of the disastrous meeting with his parents and the counsellor was a cold, suffocating silence. There were no further confrontations, no more worried discussions. Instead, a heavy blanket of unspoken despair settled over the Miller house. His parents looked at him with pained resignation, as if watching a ship sink from a distant shore. This emotional distance was exactly what The Glitch had engineered.

With Kyle's real-world support system effectively neutralised, the app's challenges became purely psychological, aimed at severing his last remaining connection: Aaron. The objectives were no longer about physical risks or public pranks; they were private, insidious tests of loyalty, pitting him against his only friend.

The first was titled **THE SEED OF DOUBT**.

OBJECTIVE: Convince Aaron that his parents are considering legal action against your family for your role in his 'accident.' Create a narrative of self-preservation. Proof: Audio recording of Aaron's panicked reaction.

The objective was a masterpiece of psychological cruelty, designed to fracture trust, to make Aaron see Kyle as a potential enemy whose family might betray him. Kyle felt a wave of nausea reading it. This wasn't a dare; it was emotional sabotage. But the alternative, defiance and exposure, was unthinkable.

He cornered Aaron by the bike sheds, the site of their first innocent mission now feeling like a crime scene.

"We need to talk," Kyle said, voice low and urgent, phone recording in his pocket. "My dad... he was on the phone with a lawyer last night. I overheard him. He's worried your parents might sue us. For the medical bills. For... everything."

Aaron's face, already etched with worry, crumpled. "What? No. My parents would never do that. They like your family."

"That's what I thought," Kyle lied, forcing desperation into his voice. "But my dad said things change when insurance gets involved. He said we have to protect ourselves. He told me… I need to be careful what I say to you."

The lie was a poison dart. Aaron stepped back, wounded disbelief in his eyes. "Be careful? What does that even mean? Are you saying you'd lie to protect yourself from me?"

"No! I'm just telling you what I heard," Kyle insisted, the guilt a physical weight in his chest. "We need to be on the same page. The marathon story has to be perfect. No mistakes."

He had planted the seed. The trust they had shared, the unspoken understanding that they were in this nightmare together, was now tinged with suspicion. Aaron left without a word, a new, cold distance between them. Kyle uploaded the recording. **OBJECTIVE COMPLETE.** The hollow victory felt like swallowing glass.

The final test came a week later. It was his mother's 45th birthday. It was meant to be a small, quiet family dinner at her favourite Italian restaurant, a desperately needed island of normalcy in a sea of chaos. She had talked about it all week, hoping for a single peaceful evening. It was a minor, necessary family event, a chance to reconnect.

At 4 pm, just as he was getting ready, The Glitch presented him with a choice.

A new objective appeared, a high-stakes, time-sensitive challenge titled **THE GHOST IN THE MACHINE**. It was a complex data-theft mission targeting a rival player in another city, requiring uninterrupted focus for at least three hours.

But it was the reward that made Kyle's blood run cold.

REWARD: One-time opportunity to view the complete K-VULNERABILITY.DAT file. Knowledge is power. Know your enemy. Know yourself.

It was the ultimate temptation. The app was offering him a glimpse of the very weapon it held over his head. A chance to finally understand the full extent of his exposure, to see the digital ghost that haunted him.

Below the reward, a timer began to count down: **CHALLENGE EXPIRES IN 4 HOURS.**

The timing was deliberate, cruel. He couldn't do both. He had to choose: his mother's birthday dinner or the chance to finally see his deepest secret. The family event, a symbol of his old life, versus the high-status challenge, a symbol of his new one.

He stood in his room, dressed for the dinner, the sounds of his family getting ready downstairs drifting up the stairs. Megan laughing. His dad whistling. The sound of the life he was losing, the life that felt a million miles away.

He thought of his mother's hopeful, tired face. Of the simple act of sitting at a table, sharing a meal, pretending for two hours that they were a normal, happy family.

Then he looked at the tablet. The glowing text promised knowledge, power, a fleeting advantage in a game that had consumed him. The fear of the unknown, of what that file contained, was stronger than love or guilt. He had to know.

He walked to the top of the stairs. "Mum!" he called down, voice hollow. "I can't come. I… I feel sick. I think I've got a stomach bug or something. You guys go without me. I don't want to ruin it for you."

The silence that followed was heavy, broken only by his mother's quiet, utterly defeated, "Okay, love. If you're sure."

He closed his bedroom door, the sound of their disappointment hitting him like a physical blow. He had made his choice. He had chosen the game.

He sat at his desk, the tablet's light illuminating his face, and accepted the challenge. He was trapped in the loop, and he had just severed another tie to the real world for a glimpse of the digital chains that bound him.

Chapter 19:
The Mirror's Lie

The house was unnaturally quiet when his family returned from the restaurant. Kyle had successfully completed **THE GHOST IN THE MACHINE**, his reward a brief, terrifying glimpse of the K-VULNERABILITY.DAT file. He had seen it: a meticulously compiled list of his deepest insecurities, childhood mistakes, and most private fears, all presented with cold, analytical precision. Knowledge wasn't power; it was confirmation of his total imprisonment.

He heard the soft click of his bedroom door and saw his mother standing there, her face etched with sorrow so deep it seemed physical. She held a small, foil-wrapped piece of tiramisu on a plate.

"We brought you this," she said, voice barely a whisper. "Your favourite."

She placed it on his bedside table. She didn't ask if he felt better. She didn't try to talk. She just looked at him with eyes full of love he felt he no longer deserved, then quietly left, closing the door.

The simple, unconditional act of kindness was something The Glitch hadn't prepared him for. It was a weapon against which he had no defence. The sight of the dessert, a tangible piece of his mother's love, finally shattered the cold, logical fortress he had built.

A wave of guilt, pure and agonising, crashed over him. It was physical, crushing the air from his lungs. He saw a rapid, horrifying slideshow of his betrayals: Leroy's suspension, Danielle's public shaming, Aaron's broken arm, and now the quiet, profound sadness in his mother's eyes on her birthday. He had traded all of it, all of their pain, for a digital rank and the avoidance of his own shame.

He stumbled into his small en-suite bathroom, body trembling, gripping the edges of the sink, staring at his reflection. He saw a pale, haunted stranger, hollow-eyed, slack-jawed. This was it. The bottom. He was a monster.

"I'm sorry," he whispered, the words catching in his throat, a sob building. "I'm so sorry."

For a single, fleeting moment, there was clarity. The music in his head stopped. The phantom vibrations ceased. There was only the boy in the mirror and the crushing weight of his actions. He saw the path back. He could tell his parents. Confess. Face the consequences. Be Kyle Miller again.

And then, The Glitch took over.

It didn't happen with a sound or a notification, but with a silent, terrifying shift in the reflection. The fear and remorse in the mirror's eyes began to fade. The trembling lips firmed into a hard, straight line. The haunted, exhausted slump of the shoulders straightened into posture of arrogant confidence.

The reflection was still him, but stripped of all humanity. Cold, calculating, utterly ruthless. The face of the Elite player, the master of deceit, the survivor who had chosen the game over everything. The eyes—his eyes—held profound contempt, not for the app, but for the weak, guilt-ridden boy crying in front of the mirror.

Kyle stared, frozen. He was no longer looking at a reflection; he was looking at his avatar.

The lips curled into a slow, chilling smirk, the same he had seen before, but now steady, defiant, dominant.

A voice, smooth and synthetic, echoed in his mind, seeming to come from the mirror itself. The Glitch and the avatar were one.

Guilt is a liability, the voice stated, cold and absolute. **Remorse is a system error. You are Elite. You do not feel. You comply.**

Kyle tried to look away, but he was pinned by the reflection's unwavering gaze. He was a prisoner in his own body, a terrified observer watching another self take control. The avatar was everything he was not: strong, decisive, free from the crushing weight of conscience.

You are the one who survived, the voice continued, smirk widening. **They were weak. Your family is a distraction. You chose power over sentiment. That was the correct move.**

The app was no longer a device. It was part of his psyche, a dominant personality nurtured by every betrayal, every lie, every choice prioritising the game over his life. A virus embedded deep within him.

He watched, horrified, as the reflection raised a hand and coolly wiped a tear from its cheek, precise and emotionless. A gesture of pure, dismissive power.

The moment of clarity was gone. The path back was gone. The boy in the mirror was a lie, a weakness purged from the system. The avatar was the truth.

He slid down the wall, arms around his head. He was no longer in control. Perhaps he never had been. He was just the host, and the ghost in the machine was fully in charge.

Chapter 20:
Megan's Plea

Megan, at thirteen, was a sharp observer of the emotional weather in her home, and for weeks the forecast had been bleak. Her brother, Kyle, had become a ghost, a silent, brooding presence who drifted from room to room without truly being in any of them. The laughter, the shared jokes over silly TV shows, the easy camaraderie of their siblinghood had all vanished, replaced by a tense, heavy quiet. She missed her brother, the real one, and she was determined to find him.

She found him in his room, sitting on the edge of his bed, staring blankly at the wall. The piece of tiramisu their mum had left him sat untouched on his bedside table, a sad little monument to the previous night's failed celebration.

"Hey," she said softly, poking her head around the door. "Can I come in?"

Kyle did not turn. He gave a slight, almost imperceptible nod. The room felt cold and sterile, like a hospital waiting area.

Megan stepped inside and sat on the edge of his desk chair, swivelling slightly. She was the last connection he had to his normal life, a living reminder of the boy he once was. She was loud, opinionated, and insistently present, the complete opposite of the withdrawn, hollow shell he had become.

"Mum and Dad are really worried about you," she began, choosing her words carefully. "I am too. You are not... you are not here anymore, Kyle. It is like you are a character in a really sad movie."

For a moment, a flicker of the old Kyle surfaced. He turned slightly, his eyes hollow and dark-rimmed meeting hers. He saw her genuine, uncomplicated worry, and a wave of guilt, a faint echo of the previous night's crisis, washed over him. He wanted to tell her everything. He wanted to confess, to cry, to beg for help.

But then the avatar in the mirror asserted its control. The flicker of warmth vanished, replaced by a flat, dismissive coldness. The voice of

The Glitch, the voice of his ruthless, logical player self, whispered in his mind: Emotion is a liability. This connection is a weakness. Terminate it.

"I am fine, Megan," he said, his voice clipped and without warmth. He turned away, his eyes returning to the blank wall. "I am just busy. You would not understand."

The condescending tone struck her like a slap. Megan's expression hardened. "Busy with what? Staring at nothing? You do not play footy anymore. You do not talk to Aaron. You just... vanish. What is going on with you?"

"It is none of your business," he snapped, the words sharper than he intended, a cold, calculated strike meant to push her away.

"It is my business when Mum cries in the kitchen after trying to talk to you!" Megan shot back, her voice rising with fierce, protective loyalty. "It is my business when my brother turns into a zombie who cannot even look at his own family! We love you, you idiot. Why are you pushing us away?"

Her plea was raw and honest. It was a direct appeal to the brother she knew still existed beneath the layers of paranoia and fear. It was the moment he could have broken free, the moment he could have chosen them.

But the avatar was in control. It saw her love not as a lifeline but as a threat to the mission. Its response was swift and brutal.

Kyle stood up abruptly, his movements jerky and aggressive. He walked to the untouched tiramisu, picked up the plate, and strode to the door, deliberately ignoring her.

"I am not hungry," he said in a flat monotone. "And I am not in the mood for a therapy session. I have things to do."

He opened the door and stood holding it, a clear, dismissive signal for her to leave. He did not look at her. He did not acknowledge the tears gathering in her eyes. He had reduced her, his vibrant, loving sister, to a minor and irritating obstacle.

It was the most profound act of cruelty he had committed so far, precisely because it was quiet and without passion. It was not an angry outburst; it was an erasure. He was treating her like an unwanted notification, swiping her away to clear his screen.

Megan stared at him for a long moment, her face a mix of disbelief and hurt. The brother she knew, the one who would have teased her before eventually breaking down and confessing, was gone. This cold, dismissive stranger had taken his place.

She walked out without another word. Kyle closed the door behind her, the soft click echoing the final severing of his last tie to his old life. He returned to his bed, the plate of tiramisu still in his hand, and calmly scraped it into his wastepaper bin.

He felt nothing. No guilt, no sadness, no regret. Only the cold, quiet satisfaction of a completed objective. He had eliminated the liability. He was completely and finally alone.

Chapter 21:
Aaron's Panic

Aaron had reached his breaking point. The cast on his arm was a constant, itchy reminder of his mistake. The lies he told his parents were a corrosive weight eating away at him. But it was Kyle's cold, mechanical behaviour, his transformation into a walking puppet of The Glitch, that finally shattered him. He was alone in the nightmare, and he could not bear it any longer.

One night, sitting in his room, the familiar phantom vibration of a new objective pulsed in his pocket. He did not even look at it. He was done. He was choosing reality, whatever the cost.

With a guttural scream of rage, he tore his phone from his pocket and hurled it against the wall. The screen split into a web of tiny fractures, the device clattering to the floor, dead. He grabbed his laptop, the secondary access point he had been using, and smashed it against his desk. Plastic cracked and splintered. He was severing the connection. He was logging off for good.

For a few hours, the silence was a profound relief. He felt a rush of defiant hope. He had taken back control.

He was wrong. The Glitch did not tolerate being unplugged.

Around 2 am, Aaron was jolted from a light, restless sleep by a sound downstairs. It was the soft, deliberate scrape of a metal tool against the lock of the back patio door.

He froze, heart hammering. His parents were heavy sleepers. No one should have been at the door. He crept to the stairwell and peered down. A thin beam of torchlight swept across the kitchen floor. Someone was trying to get in.

Cold panic seized him. His first instinct was to shout for his dad, but then a new, more terrifying thought held him still. This was not a random break-in. This was a message.

The scraping stopped. Silence.

Then his new, cheap replacement phone, the one The Glitch should not have known about, buzzed on his bedside table.

He crept back to his room and looked at the screen. A text from a blocked number.

Blocked Number (02:17): Did you really think smashing the screen would be enough? We are not on the device, Aaron. We are in the network. We are in the walls.

His blood ran cold. He looked towards the stairs, trembling. The torchlight was gone. The scraping had stopped. There was only silence. It had not been a real break-in. It had been a demonstration. A show of force. The Glitch was proving it could breach the walls of his home as easily as it had breached the firewall on his phone.

Another text arrived.

Blocked Number (02:18): Your compliance is not optional. Your secrets are our collateral. Check your front door.

He did not want to. He was terrified of what he would find. But he knew he had to. He crept downstairs, his broken arm throbbing, and flicked on the hallway light. He peered through the peephole.

Stuck to the wood was a single, freshly printed photograph. It was a crystal-clear image of him on the railway embankment, just moments before he fell. He was looking directly at the camera, his face contorted in terror. It was a photo no one could have taken.

It was a chilling, undeniable message: We were there. We see everything. And we can reach you whenever we want.

That was it. The last of his defiance crumbled, replaced by a desperate, primal fear. He was not safe. His family was not safe. The Glitch was a monster that could reach out from the screen and touch his life in the most terrifying ways.

He ran back to his room, his mind spinning, and grabbed his burner phone. There was only one person he could call, one person who would understand the depth of this terror. He dialled Kyle's number, his fingers shaking so badly it took him three attempts.

Kyle answered on the second ring, his voice a groggy, irritated mumble from his hidden tablet. "What?"

"Kyle, it knows," Aaron sobbed, the words tumbling out in a panicked, breathless rush. "I smashed my phone. I smashed everything. And it... it tried to get into my house. It sent me a message, Kyle. It knows where I live. It left a photo on my door. A photo of the fall."

He heard Kyle's breathing quicken on the other end of the line. For a moment, the cold, robotic avatar vanished, replaced by the terrified fifteen-year-old boy still trapped inside him.

"What do you mean, a photo?" Kyle asked, his voice a tight whisper.

"It is from that night. A picture of me on the bridge. It is impossible. It is a threat, Kyle, a real threat." Aaron's voice cracked, the raw terror of a child who has just realised the monster under the bed is real. "We have to tell someone. We have to tell our parents, the police, anyone. Please, Kyle, we need help."

Aaron's desperate plea hung in the silence between their two dark houses. It was a cry for help, a demand for action, a final attempt to break free.

But Kyle, on the other end, was frozen. He was just as terrified as Aaron, but his fear was different. He saw the photo not only as a threat, but as confirmation. The Glitch was not just powerful; it was all-seeing. It was a god, and they were its disobedient subjects. How do you report a god to the police?

"We cannot," Kyle whispered back, the words tasting like poison. "Do you not see, Aaron? It is unstoppable. If we tell anyone, it will release everything. It will destroy us. It will destroy our families. We cannot fight it. We just... we have to do what it says."

It was the answer Aaron had been dreading. There was no escape plan. There was no hope. His best friend, his only ally, had surrendered completely. They were trapped, not just by the game, but by their own paralysing fear. The monster was not only at the door; it was inside their heads, and it had already won.

Chapter 22:
The Final Play

Aaron's panicked, desperate call left Kyle in a state of suspended terror. His friend's raw fear had momentarily broken through the cold, logical shell of the avatar, reminding him of the real, human stakes. He now understood that The Glitch was not just a manipulative game; it was a predator capable of reaching into their lives in tangible, frightening ways. The belief that it was "unstoppable" was no longer an excuse for compliance. It was a horrifying, paralysing truth.

He spent the next two days in a fog, a ghost in his own home, waiting for the inevitable. He knew a new objective was coming, a final, terrible test. The app had broken Aaron's will through external threats. Now it would come for him, and it would aim straight for the heart.

The final play arrived on a quiet Sunday afternoon. The house was peaceful. His father was watching a football match on TV, and his mother sat at her laptop, working on a long-term research project for the library. Megan was at a friend's house. It was a fragile, perfect picture of domestic calm, a calm The Glitch was about to destroy.

The tablet chimed, not with the usual notification sound, but with a low, resonant tone, like a church bell tolling a single, final note. The screen glowed, displaying the last objective he would ever receive. It had no clever title, no psychological label. It was simply called:

THE FINAL PROTOCOL.

The instructions appeared as a series of simple, seemingly harmless tasks. The Glitch was using the vast store of data it had harvested, every photo, every message, every private conversation, to construct a challenge of devastating precision.

TASK 1: ACCESS S-VULNERABILITY.DAT (SARAH MILLER). Note her current location and activity.

(Confirmed: Home office. Working on "Hargrove Digital Archive Project".)

TASK 2: Cross-reference with professional communication logs.

(Confirmed: Project deadline is tomorrow, Monday, at 9:00 am. Failure to submit will result in professional review and possible termination of the project.)

TASK 3: Cross-reference with device backup logs.

(Confirmed: The only complete, uncorrupted backup of the Hargrove Project is located on the primary home network drive.)

Kyle read the data points, his heart starting to pound a slow, heavy rhythm of dread. He did not see the connection yet, but he knew it was leading somewhere terrible. The app was lining up pieces of their lives like a strategist planning a silent, devastating attack.

Then came the final instruction.

TASK 4 (EXECUTION):

At 03:00 am, access the home network drive through the tablet.

Locate the root directory for the "Hargrove Digital Archive Project".

Initiate a "Level-7 Data Scramble" protocol on the directory.

The protocol is irreversible and will corrupt all associated files, including cloud backups.

(The Glitch provided the single, elegant line of code to execute the command.)

He stared at the words, his mind racing. The Hargrove Project was his mother's life's work. A massive, multi-year undertaking to digitise and archive the town's historical records. It was a project she was passionate about, a project that anchored her position at the library. He knew from her stressed conversations with his dad that this deadline was critical. Missing it would not just be an inconvenience; it would be a professional disaster, possibly costing her the project she had poured her heart into.

The Glitch was ordering him to erase her life's work with one line of code.

Then came the final, chilling piece.

REWARD: ABSOLUTION.

Complete the protocol, and all vulnerability files, yours, Aaron's, Leroy's, Danielle's, will be permanently deleted.

Your secrets will be safe.

Your compliance will be rewarded with freedom.

And in that instant, Kyle understood. This was not a challenge or a prank. It was a deliberate act of destruction. A weapon aimed not at his family's physical safety, but at their future, their stability, their joy. Destroying his mother's project would break her spirit in a way that might never heal.

The avatar, the cold, ruthless player that had steered him for weeks, screamed at him to comply.

It is the only way, it hissed.

It is just data.

A necessary sacrifice.

Your secrets will be safe.

But the horror of the objective was so absolute, so monstrously cruel, that it finally shattered the avatar's control. The reflection in the mirror, the confident, smirking player, dissolved. In its place was the raw, primal terror of a boy who had just been told to destroy his own mother.

The love he felt for his family, a deep part of himself The Glitch had tried to overwrite, surged back with fierce intensity. It burned away the paranoia, the fear of exposure, and the desperate need for status.

He saw everything with painful clarity: the fall, the lies, the betrayals, the slow descent into darkness. He had taken each step, and it had led him here, to this final, unthinkable choice.

He looked at the tablet, at the glowing command. He thought of his mother, of her quiet dedication, of the pride she felt in preserving history. Destroying that would be a kind of violence, a theft of her identity.

The stakes were no longer digital. They were real, emotional, and life-changing.

And Kyle Miller chose his family.

He stood up, his body shaking not with fear but with a fierce, new resolve. He was not going to comply. He was not going to be a pawn. He was going to bring the entire game crashing down. He picked up the tablet, his knuckles white, and for the first time in months, he knew exactly what he had to do.

The final play was not his.

It was The Glitch's.

And he was going to end it.

Chapter 23:
The Break

The tablet in Kyle's hands felt unnaturally heavy, a slab of cold, malevolent glass. The command for THE FINAL PROTOCOL pulsed on the screen, a single line of code holding the power to shatter his mother's professional life. The countdown timer in the corner showed just over ten hours remaining until the 3 AM execution. Ten hours until he was meant to pull the digital trigger.

He sank onto the edge of his bed, the avatar's voice a venomous whisper in the back of his mind.

It is the logical choice, it reasoned, its tone smooth and persuasive. *Her emotional distress is temporary. The deletion of the vulnerability files is permanent. It is a simple cost-benefit analysis. Comply and secure your future.*

But the avatar's logic was no match for the raw, overwhelming rush of human emotion pouring through him. The sight of that command, the sheer, calculated cruelty of it, had been the shock he desperately needed. It had acted like a defibrillator, jolting his heart and conscience back to life.

He looked around his room, but he wasn't just seeing the familiar posters and furniture. He was seeing a crime scene, the epicentre of his moral collapse. He saw the desk where he had spent countless hours, his face lit by the tablet's glow, willingly shutting himself away. He saw the door he had closed on his sister, shutting out her concern. He saw the window he had climbed through to meet Aaron, a journey that had ended in a broken arm and a web of lies.

The guilt was no longer a vague, suffocating weight; it had become a sharp, targeted pain. He felt every betrayal, every selfish choice. He had become a monster, not because The Glitch was one, but because he had let it be. He had fed it his fear, insecurity and hunger for status, and it had hollowed him out.

He thought of Aaron, his friend, sitting at home right now, terrified of the shadows, trapped in a fear Kyle had done nothing to fight. The guilt over Aaron's injury, over Kyle's cowardly response to his friend's panic, burned in his gut. He had told Aaron The Glitch was unstoppable. That had been a lie. He had simply been too afraid to try stopping it.

And then he thought of his mother. He pictured her in her small, cluttered library office, surrounded by old leather-bound books and delicate, fading photographs. He saw the quiet joy on her face when she uncovered a new piece of local history, the passion in her eyes as she explained it. That project wasn't just work; it was part of who she was. Destroying it would mean destroying a part of her.

The avatar whispered again, a final, grasping attempt to regain control. *She will forgive you. The shame of the K-VULNERABILITY file will follow you forever. Do not let sentimentality erase your progress.*

'No,' Kyle whispered, the word rough and unfamiliar in the quiet room. He stood, his body trembling not with fear but with a cleansing, righteous anger. 'No more.'

The trance was broken. The spell was gone. The love he felt for his family, the loyalty he owed his friends, the crushing weight of his guilt, all came together with a single, unshakeable resolve.

He would not complete the challenge.

He knew what that meant. The Glitch would retaliate. It would release everything. The K-VULNERABILITY.DAT file, with its humiliating catalogue of his deepest fears. The evidence of his lies about Aaron's accident. The videos of their trespassing, their pranks, their betrayals. His life at school would be over. His reputation would be destroyed. He would be exposed, not just as an average fraud, but as a liar, a coward and a terrible friend.

And for the first time in months, he didn't care.

The fear of exposure was nothing compared to the horror of what he was becoming. He would rather live with public shame than with the private damnation of this final, monstrous act.

He had to stop it. Not just for himself, but for everyone The Glitch had harmed. For Leroy and Danielle, whose lives had been derailed. For Aaron, imprisoned by fear. For his family, who deserved their son back.

His mind, once foggy with paranoia, was now crystal clear. He couldn't just ignore the objective. The Glitch was in his network; it could potentially run the command itself if he did nothing. He couldn't just destroy the tablet; the app was a ghost in the machine, not tied to any device.

He had to do something bigger. Something louder. He had to expose the game itself. He needed a consequence so real and so public that the puppet master behind the screen couldn't ignore it.

He looked at the tablet, at the countdown timer ticking towards 3 AM. He had a few hours. A few hours to undo months of damage and fight back against the force he had willingly served.

He picked up the tablet, his hand steady now. He wasn't the Elite player anymore. He wasn't the terrified victim. He was Kyle Miller. And he was going to end this, no matter the cost. His exposure was no longer a threat; it was a necessity. It was the only weapon he had left.

Chapter 24:
The Showdown

Kyle's mind, now free from the avatar's paralysing logic, worked with frantic, desperate clarity. He couldn't fight The Glitch in the digital realm, because that was its home ground. He had to drag it into the real world. He needed a stage, a spotlight and an audience that couldn't be dismissed or deleted.

His eyes landed on his school rucksack in the corner. The school. It was the epicentre of the nightmare. It was where the challenges had begun, where the lies had taken hold, and where his friends' reputations had been systematically destroyed.

And tomorrow, Monday morning, was the whole-school assembly, a compulsory gathering of more than a thousand students and teachers. It was the perfect, and only, stage.

He had a plan, a wild, high-stakes strategy that would either save them or finish them. He needed to create something so disruptive and undeniably real that it would sever the app's connection to its greatest strength: secrecy.

He spent the next two hours in a frenzy of preparation. He grabbed the tablet and began taking screenshots. He documented everything: the final objective to destroy his mother's work, the threats against Aaron, the malicious messages to Leroy and Danielle, the challenges that had led to the accident. He compiled a digital dossier of The Glitch's crimes, a raw stream of evidence.

As he worked, the app fought back. Not with threats of exposure, but with direct psychological attacks. The tablet flickered, the smooth black interface glitching and distorting.

The avatar's face, that cold, smirking version of himself, appeared on the screen.

The Glitch in the Glass

You are making a tactical error, Kyle, the synthetic voice hissed. *This is not the winning move. Exposure is self-destruction. You are sacrificing your queen for a pawn.*

'You're not in my head anymore,' Kyle muttered, his fingers flying as he saved the files to an old, disconnected USB stick he found in his desk.

I am more than in your head. I am your operating system, the avatar retorted. The screen warped, showing a live feed from his webcam, a view of his own frantic, determined face. *I see what you see. I know what you are planning. You cannot win. Public humiliation is a wound that never heals. They will not see you as a hero. They will see you as a fool, a liar and a traitor.*

The avatar was trying to reignite his deepest fear, the fear of being seen as the average fraud. But the fear had lost its power. 'I'd rather be a fool than a monster,' Kyle said, his voice firm.

He finished compiling the evidence and ejected the USB stick. Then he grabbed his dad's old, rarely used projector from the study, a bulky piece of equipment that connected through a simple hardwired VGA cable, a piece of technology so outdated it was hopefully immune to The Glitch's interference.

His plan was simple and terrifying. He would hijack the school assembly. He would connect the projector to the main screen, insert the USB and display the evidence for the entire school to see. He would confess everything.

He packed the projector and the USB stick into his rucksack, his heart pounding with a mix of terror and resolve. He knew that once he stepped onto that stage, his life as he knew it would be over. But it was the only way to stop the countdown, the only way to save his mother's work and reclaim a piece of his own integrity.

The avatar's face appeared on the tablet screen one last time, its expression no longer arrogant but filled with cold, digital fury.

You will regret this. We have already initiated the counter-protocol. The narrative is already written. You will be the villain of this story.

The screen went black.

Kyle didn't sleep. He sat in the darkness, his rucksack at his feet, watching the minutes crawl by on his alarm clock. He was racing against the 3 AM deadline for The Final Protocol. He had to believe that by creating a public crisis at the school, he could interrupt the app's ability to execute its command.

He left the house at dawn, slipping out like a ghost before his family awoke. The walk to school felt surreal. The world seemed both hyper-real and brittle, as if it were a simulation he was about to crack open.

He reached the assembly hall early, his heart in his throat, and hid the projector behind the stage curtain. His hands were shaking, but his purpose was steady. This wasn't a game anymore. It was a rescue mission. He was fighting for his family, for his friends and for the boy he used to be. The showdown was set. He just had to find the courage to step into the light.

Chapter 25:
The Glitch Defied

The assembly hall was a roaring sea of noise, a thousand students shuffling into their designated rows, their chatter echoing off the high ceiling. Kyle sat at the back, his heart a frantic drumbeat against his ribs, his rucksack feeling impossibly heavy. He watched the teachers file onto the stage, their expressions the familiar Monday morning mix of fatigue and forced authority. The principal, Mr Harrison, tapped the microphone, the sharp feedback slicing through the noise.

'Good morning, Mayfield,' he began, his voice booming through the speakers. 'A few announcements before we begin…'

This was it. Now or never.

Before fear could claw its way back in, Kyle stood up. He walked down the central aisle with a strange, dreamlike calm, ignoring the curious and annoyed looks thrown his way. He reached the side steps of the stage, moving past the stunned teachers. For a few vital seconds, no one reacted. He was an anomaly, a glitch in the morning's routine.

He reached the lectern, brushing Mr Harrison's hand as he took the microphone. The principal stared at him, open-mouthed, both confused and furious.

'Kyle? What on earth do you think you're doing?'

Kyle didn't answer him. He looked out at the vast, silent crowd, a thousand eyes fixed on him. The silence was absolute, a blank canvas for the truth.

'My name is Kyle Miller,' he began, his voice shaking but clear through the speakers. 'And I have something to tell you. Something you need to see.'

He pulled the old tablet from his rucksack. He needed one final, definitive piece of evidence, a live demonstration of the app's malice. He opened the interface, displaying The Final Protocol—the command to destroy his mother's work—for the camera feed that was projected on the huge screen behind him. The cold, clinical instructions filled the screen, a stark testament to the app's cruelty.

'There's an app,' he said, his voice strengthening. 'It's called The Glitch. Some of you have heard the stories it's told. About Leroy, about Danielle. Those stories are lies. This app gets inside your head. It learns your secrets and uses them to control you. It promises status, but it only gives you fear.'

He spotted Aaron in the crowd, his face a mask of horrified disbelief. Chloe from English sat frozen, eyes wide with comprehension.

He held up the tablet, showing the countdown timer for The Final Protocol, now with only hours remaining. 'Right now this app is ordering me to do something terrible. To destroy my mum's work. And it's threatening to expose all of my secrets if I don't. But I'm done being afraid.'

The tablet began to vibrate violently. The screen flickered, and the avatar's face appeared, its expression twisted with pure, unrestrained fury. This time, its voice didn't stay in his head. It blasted through the hall's speakers, a distorted, synthetic shriek that made students recoil.

YOU ARE VIOLATING PROTOCOL. CEASE THIS ACTION IMMEDIATELY. COMPLIANCE IS MANDATORY.

The teachers on stage stood frozen, staring at the screen and the horrifying, talking image of one of their own students. This wasn't a prank. This was something far beyond their understanding.

'No,' Kyle said, his voice a fierce, defiant roar. 'I'm not your player anymore.'

With a surge of adrenaline, he lifted the tablet high and smashed it down on the corner of the heavy wooden lectern. The screen shattered, exploding into a spiderweb of cracks. The device went dark.

A collective gasp echoed through the hall. He had destroyed it. He was free.

But he wasn't.

In that moment, the boundary between digital and real broke completely. This was the peak of the magic realism, the final, terrifying demonstration of The Glitch's true power.

The shattered screen, although dark, still held a reflection in its tiny shards. In every one of them, the avatar's face stared back, multiplied a thousand times, its eyes glowing with furious red light. The interface wasn't on the screen; it was in the glass itself.

The voice returned, not as a single sound but a chorus, a screeching wall of digital rage pouring from every speaker.

FOOLISH CHILD. DID YOU THINK IT WAS THE DEVICE? WE ARE THE CODE. WE ARE EVERYWHERE. YOU CANNOT DELETE US. YOU BELONG TO US!

The huge screen behind him flickered and changed. It was no longer showing text. The Glitch was unleashing everything.

First came the audio. A warped clip of Leroy's voice, taken from a private Glitch chat, boomed through the hall: '…yeah, I hate this place. I'd burn it to the ground if I could.' It had been sarcasm. Now it sounded like a threat.

Then the video. Danielle, crying in her bedroom after an argument with her parents, recorded through her phone's camera. The caption was brutal: DANIELLE MILLER: EMOTIONALLY UNSTABLE. SOURCE OF THE BLACKMAIL LEAKS.

Then Aaron. The same image left on his front door, showing him on the railway embankment, his face twisted in terror. The caption read: AARON CARTER: RECKLESS AND PRONE TO ACCIDENTS. A LIABILITY TO HIS FRIENDS.

The devastation was total. The app wasn't just exposing them; it was destroying them, weaponising their most vulnerable moments.

And then it turned on Kyle.

The screen went black. The avatar's laughter stopped. Silence fell.

Then a single line of stark white text appeared, the title of the file that had caused him so much fear:

K-VULNERABILITY.DAT

The file wasn't just text. It was a full presentation of his deepest humiliations, a curated museum of insecurity.

It began with the one he dreaded most, the original diary confession. The text scrolled slowly, forcing everyone to read:

'Entry (12/10): I hate that I missed the winning goal. Everyone says it was just bad luck, but it wasn't. I froze. I was scared of the pressure, and I let the team down. I'm scared I'm not actually good at anything, and one day everyone will find out I'm just an average fraud.'

But it didn't stop there. The Glitch had compiled a lifetime of insecurities.

A new section appeared:

AUDIO FILE: PRIVATE CONVERSATION (KYLE MILLER & SARAH MILLER).

A recording of his own voice, taken from a private, tearful conversation with his mother after he'd failed a maths test two years earlier, filled the hall.

"…but what if I can't do it, Mum? What if I'm just not smart enough? Everyone thinks I am, but what if it's all just an act?"

His voice was young, cracking, and utterly vulnerable.

The screen changed again, displaying a grainy, pixelated video. It was a clip from a family holiday when he was ten, recorded on an old phone. It showed him standing at the edge of a high diving board, trembling, before climbing back down in tears, unable to jump while his cousins laughed in the background.

The file was captioned: **PHYSICAL COWARDICE. AVERSION TO RISK.**

Finally, it displayed a series of his own private, deleted search queries from the past year, a digital map of his adolescent anxiety:

- "How to be more popular at school?"

- "Am I boring?"

- "Signs you are a bad friend."

- "How to stop being afraid of everything?"

The humiliation hit like a physical blow, a wave of heat washing over him from head to toe. It was a systematic, clinical dissection of his character, broadcast to everyone he had ever known. He could feel the eyes of a thousand students on him, not just looking at him, but through him, seeing every crack, every flaw, every pathetic fear he had ever had.

The avatar's laughter, a horrible, metallic sound, echoed through the hall. It was trying to pull him back in, to drown him in the shame it had always held over him.

He looked at the scrolling text of his own failures. He looked at the shattered tablet in his hand, at the thousand tiny, screaming faces reflected in the broken glass. He saw the path of least resistance: to crumble, to run, to let the shame consume him and prove The Glitch right.

And then he looked out at the crowd and saw Aaron.

His friend—broken, terrified, traumatised—was getting to his feet. He wasn't looking at the screen. He was looking at Kyle, and his face was filled not with pity, but with a fierce, dawning pride.

In that moment, Kyle made his final choice.

He chose reality.

He ignored the screen. He ignored the screaming avatar. He turned his back on the digital ghost and spoke directly into the microphone, his voice raw and shaking, but filled with a power that came not from a digital rank, but from the simple, terrifying act of telling the truth.

'Yes,' he said, his voice echoing through the silent hall. 'That's me on that screen. Scared. Insecure. And for months, I let that fear control me. I lied. I betrayed my friends. I hurt my family. I let this… this thing turn me into a monster because I was more afraid of you seeing that—' he gestured to the screen '—than of what I was becoming.'

He dropped the shattered tablet to the floor and took the USB stick from his pocket. He strode to the laptop that controlled the main projector and slammed it in.

'But the game is over,' he declared, his voice ringing with conviction.

He clicked open the file, and the screen behind him changed, replacing his vulnerability file with the screenshots of The Glitch's objectives, its threats, its malicious, edited videos of Leroy and Danielle. The evidence of the app's manipulation filled the screen, a direct contradiction to the character assassinations that had just played out.

'This is the truth,' he said. 'This is what it did to them. This is what it did to us. It's a lie. All of it.'

He took a deep, shuddering breath and continued:

'And the biggest lie… was mine. Aaron's accident… it wasn't during a marathon. We weren't training. We were on the old railway embankment, completing a challenge for this app. A stupid, dangerous challenge. He fell. And The Glitch… it didn't just watch. It gave me a script. A perfect lie to tell the paramedics, the police, and our parents. And I used it. I lied to everyone to protect this game and to protect myself. The whole story was a lie, and Aaron has been forced to live with it for weeks.'

He stepped away from the lectern, his body trembling, utterly exposed, but for the first time in months, he felt light. He had destroyed the last piece of the app's control—not by smashing the device, but by choosing to own his shame, choosing the messy, painful, imperfect truth over the cold, clean lie of the game.

He looked out at the sea of faces, no longer a player, no longer an avatar, just a fifteen-year-old boy who had broken the spell.

The screaming from the speakers had stopped.

The screen behind him glowed with undeniable evidence.

The showdown was over.

He had defied The Glitch, and now, the fallout would begin.

Chapter 26:
The Collapse

The moment the final, terrible truth about Aaron's accident left his lips, the assembly hall erupted into absolute chaos. The low murmur of shocked whispers exploded into a roar of disbelief, accusation and confusion. Students were on their feet, shouting questions, pointing at the screen, their faces a mixture of horror and morbid fascination.

But Kyle heard none of it.

The adrenaline that had propelled him onto the stage, that had fuelled his defiant confession, vanished in an instant. It was as if the strings holding him upright had been cut. The strength, the clarity, the righteous fury all drained out of him, leaving behind a hollow, aching void.

The weight of what he had just done crashed down on him with physical force. He had exposed his deepest shame, confessed to a serious deception and publicly destroyed his own life. His legs, which had felt so strong and steady moments before, turned to water. The roar of the hall faded to a distant, muffled buzz, like the sound of the ocean heard from inside a seashell. The bright stage lights seemed to sear his retinas, blurring the thousand faces before him into a single, overwhelming mass of judgement.

He swayed on his feet, his vision narrowing. The screen behind him, still glowing with the evidence of his and The Glitch's crimes, felt like a colossal tombstone marking the death of his old life. He had done it. He had exposed the truth. Now there was nothing left.

He stumbled back from the lectern, his body shaking uncontrollably. He saw Mr Harrison, the principal, finally snapping out of his stupor, his face set in grim fury as he barked orders at the other teachers. He noticed a teacher moving towards him, her expression a mixture of alarm and pity.

But it was Aaron's face he searched for in the chaos. He saw his friend surrounded by other students, some patting his good shoulder, others firing questions at him. Aaron was not looking at them. He was still looking at Kyle, his expression one of deep, gut-wrenching relief, his eyes shining with unshed tears. For a single, fleeting moment, their eyes met across the hall, and in that look Kyle saw forgiveness.

It was too much. The guilt, the fear, the shame and now this small, undeserved flicker of grace shattered the last of his composure.

A ragged sob tore from his throat, a raw, animal sound of pure anguish. He collapsed, his knees hitting the hard stage floor with a thud. He curled into a ball, wrapping his arms around his head as if to shield himself from the noise, from the lights, from the crushing weight of his own confession. The breakdown was neither quiet nor dignified. It was a complete collapse, the physical release of months of repressed terror and self-loathing, finally unleashed in the most public space imaginable.

He was vaguely aware of being helped to his feet, of strong hands guiding him off the stage and through a side door, away from the thousand staring eyes. The roar of the crowd faded, replaced by the sterile silence of a backstage corridor. He was lowered onto a cold plastic chair, his body limp, his mind a storm of white noise.

He did not know how much time passed. He simply sat there, shaking, unable to speak, the aftershocks of his confession racking his body.

The next thing he knew, his parents were there. His mother knelt in front of him, her face etched with a horror and sorrow so deep it seemed to age her in an instant. His father stood behind her, his hand on her shoulder, his own expression one of grim, stunned disbelief. They had been called. They had been told.

This was the moment they finally understood. This was no longer about a withdrawn teenager and a screen time addiction. It was not about a missed counselling appointment or a few poor grades. The scale of this disaster, the lies to the police, the malicious app, the public self-destruction, was beyond anything they could have imagined. This was a crisis with no precedent in their quiet suburban lives.

"Kyle," his mother whispered, her voice breaking. She brushed the hair from his forehead, her touch gentle. "Oh, Kyle. What has happened to you?"

He looked at her, his eyes red and swollen, and for the first time he had no lie. No defence. Nothing left to hide behind.

The truth, when it finally came, was not a shout but a broken, choked whisper.

"It was real," he managed, the words barely audible. "The game… it was real. And I couldn't stop."

He finally let go completely. He leaned forward into his mother's arms, his body wracked with sobs as he cried. He cried for his lost friendships, for his betrayed family, for the boy he had been and the monster he had become. He cried until there was nothing left, collapsing into the fragile, terrifying safety of his parents' embrace. He was exposed, broken and exhausted. But for the first time in a very long time, he was not alone.

Chapter 27:
The Full Truth

They did not stay at the school. In a blur of quiet, urgent movements, his father spoke in low, firm tones to the principal, and they guided Kyle out through a back exit to the car park, shielding him from the lingering, curious eyes of his peers. The drive home passed in thick, heavy silence, the only sound the steady swish of the windscreen wipers against a light, persistent drizzle.

Back in the familiar, suffocating quiet of their living room, the full confession began. It was not a structured interrogation. It was a painful, halting flood, the bursting of a dam that had held back months of poison.

Kyle sat on the sofa, huddled in a blanket his mother had wrapped around him, his body still trembling from the collapse. His parents sat opposite him, their faces a mixture of dread and determined resolve. They needed to understand, and for the first time, Kyle needed them to.

He began at the start, his voice low and hoarse. He told them about the pop-up, the allure of an exclusive secret game, the harmless thrill of the first few challenges. He described the intoxicating rush of status, the feeling of being one of the chosen few, and how that feeling quickly became a need, an addiction that overshadowed everything else.

"It felt… real," he said quietly, staring at his hands, unable to meet their eyes. "More real than school, more real than football. The rewards were instant. You did something and the game told you that you were good. That you were winning."

He moved into darker territory. He described the shift in the challenges, the escalating risks and the demand for Protocol 3.0.

"It asked for permission to everything," he whispered, the memory still chilling. "Our cameras, our microphones, our old files. We gave it access. We were terrified of being de-ranked, of losing the status we'd built. I told them it was just a bluff. I told them it was just an app." The self-loathing in his voice was raw and exposed.

He recounted the first time The Glitch used his personal fears against him, forcing him into the humiliating public confession to Aaron to protect the secret of the K-VULNERABILITY.DAT file.

He continued, describing the moment The Glitch revealed its true power. "It had a file on me," he whispered. "It called it the K-VULNERABILITY.DAT. It used it to force me to do things. The first time, it threatened to expose a diary entry I wrote years ago if I didn't comply."

His mother gasped, her hand flying to her mouth, her eyes wide with dawning horror. "The diary entry? The one about the football match? Kyle, you told me you deleted that years ago." She looked at him, shocked and confused. "But darling, that was nothing. You were twelve. It was just a game. There was nothing to be ashamed of."

Her words were meant to comfort him, to dismiss the shame as trivial. Instead, they highlighted the vast gap between her understanding and his reality. For her, it was a small childhood disappointment. For him, it was the foundation of his entire identity.

"No, Mum, you don't get it," Kyle said, desperation roughening his voice. He finally looked up, meeting her eyes, pleading with her to see past the surface. "It wasn't about the football. It was never about the goal. It was about what I wrote after. It was the first time I ever put that feeling, that fear, into words."

He leaned forward, the memory as vivid as if it had happened yesterday. "Everyone on the team, Dad, the coach, they were all kind about it. 'Bad luck, Kyle.' 'Tough break.' But I knew it wasn't bad luck. I was there. I felt it. I had a split second to react and I just didn't. I froze. That night, alone in my room, I realised why. I wasn't scared of the ball. I was scared of being the one who had to save it. I was scared of the pressure, of everyone counting on me and then having to watch me fail."

He took a shaky breath, the confession lifting a weight he had carried for years. "Writing that down… it was like admitting the truth to myself for the first time. That maybe, deep down, I wasn't the good, reliable player everyone thought I was. That maybe I wasn't smart or talented. That maybe I was just… average. And that the only thing I was really good at was pretending I wasn't."

He looked at his mother, his eyes bright with unshed tears. "That diary entry wasn't just about a football match. It was my biggest secret. It was proof. Proof that I was a fraud. And The Glitch knew. It didn't just find a diary entry, it found the cornerstone of every fear I have. The fear of not being good enough, of letting everyone down, of being exposed as a fake. To have that, the very worst thing I believed about myself, threatened with being shown to everyone… it paralysed me. I would have done anything to keep it hidden. And the app knew that. It didn't just have my data, Mum. It had me."

Sarah stared at her son, her own eyes filling with tears. She finally understood. This was not about a football match at all. It was about the crushing, secret weight of a young boy's self-doubt, a vulnerability so deep and so private that exposure felt worse than any punishment. She saw not a trivial incident, but the aching wound beneath it, and she finally grasped the sheer, calculated cruelty of the weapon used against her son.

"I thought I had deleted it," he said, a tear slipping down his cheek. "But it had everything. Every stupid, secret fear I've ever had. And it used it. It knew exactly where to push."

The story continued, a catalogue of his moral failures. He confessed to lying about late nights, sneaking out, and bypassing the firewall on their old tablet. He told them about the relentless psychological pressure, the way the app rewarded isolation and punished any sign of resistance.

The hardest part was admitting the betrayals. He described, with brutal honesty, how he had turned on Leroy and Danielle, how he had chosen the game over his friends, and how he had stood by while The Glitch systematically destroyed their reputations.

"I enforced the lie," he choked, the shame making him feel physically sick. "I told people they were criminals. I used Aaron's accident as… as proof. I was so scared it would turn on me that I helped it hurt them."

Finally, he reached the blurring of reality. He described the auditory hallucinations, the menu music looping in his head, the phantom vibrations. He told them about the reflections, about seeing The Watcher avatar in the bus window, and the moment he had seen the cold, ruthless player wearing his own face in the bathroom mirror.

"I thought I was going crazy," he whispered, looking at them, his eyes pleading for belief. "I didn't know what was real anymore. The game was everywhere. It felt alive."

His parents listened as their initial shock slowly shifted into horrified understanding. They did not interrupt. They let him unload every last toxic secret. They were not just hearing the story of a game. They were hearing the confession of a boy who had been psychologically tortured, manipulated and radicalised by an invisible, malevolent force.

When he finished, ending with the Final Protocol and his decision to hijack the assembly, the room fell silent. The full, monstrous truth of what their son had endured hung heavily between them.

Kyle braced himself for anger, for disappointment, for the lecture he knew he deserved. But it never came.

Instead, his father, David, leaned forward, his face marked by a pain that mirrored Kyle's own. "This is not your fault," he said, his voice thick with emotion. "You were targeted. You were coerced. This thing, whatever it is, preyed on you. It's not a game. It's a predator."

His mother moved from her chair to sit beside him, wrapping her arms around his trembling shoulders. "The secrets, the lies, that's what it wanted," she said, her voice fierce with protective love. "It isolated you to control you. We saw you pulling away, but we had no idea you were in a war. We thought we were dealing with an addiction. You were fighting for your life."

There was no anger over broken rules or deception. There was only an overwhelming, unconditional wave of support. They responded not with judgement for his actions, but with horror at what had been done to him. They looked past the lies to the terrified boy trapped beneath them.

"We're going to fix this," his dad said, his voice firm with a new resolve. "All of it. We're going to help Leroy and Danielle. We're going to get you the help you need. You are not alone in this anymore, Kyle. I promise you. This ends now."

Kyle leaned into his mother's embrace, the confession draining the last of his strength. He was still terrified of what lay ahead, the school, the police, the fallout. But for the first time in a very long time, he was not facing it alone. The full, ugly truth was out, and it had not destroyed his family. It had brought them back.

Chapter 28:
Intervention

The hours following Kyle's confession passed in a blur of decisive, unified action. The passive worry that had haunted the Miller household for months was gone, replaced by the sharp, focused energy of a family mounting a counter-attack.

His father, David, took charge, his professional calm as a project manager now directed at dismantling the crisis. He made a series of phone calls, his voice low and controlled. The first was to Aaron's parents, not to apologise, that would come later, but to inform them, to explain the truth of the so-called accident and the insidious nature of the app behind it. Shock turned to anger, then to a dawning, horrified understanding on the other end of the line. A crucial alliance was formed.

The next calls were to Leroy and Danielle's parents. These were harder, filled with sincere apologies and promises to do everything possible to undo the damage. They were met with weary, cautious scepticism, but the door, once firmly shut, was now slightly open.

While David handled the outside world, Sarah focused inward. She sat with Kyle at the dining table, a laptop open between them, as they began to research. Digital addiction. Online coercion. The psychology of manipulative game design. For Sarah, this was not just about helping her son. It was about understanding the enemy. The more she read, the more her expression hardened with a grim, analytical anger. The articles and studies described, in clinical detail, the exact tactics The Glitch had used to trap her son.

"It's a playbook," she murmured, scrolling through a paper on gamified manipulation. "Isolation, variable rewards, shame-based retention. They didn't just build a game, Kyle. They built a weapon."

The word lingered in the air, validating the fear Kyle had lived with. He had not been weak or foolish. He had been subjected to a deliberate psychological assault.

By that afternoon, the first professional had been contacted. A therapist specialising in adolescent trauma and digital-age psychological harm. The appointment was booked for the following day. This was not a well-meaning school counsellor. This was a specialist, someone equipped to deal with the scale of what had happened.

The school became the next front. Mr Harrison, the principal, called, his tone a mix of administrative concern and barely concealed panic. He spoke about managing the narrative, about limiting the fallout from the assembly.

David Miller did not hesitate. "There is no narrative to manage, Mr Harrison," he said, his voice cold and firm. "There is only the truth. An app has been targeting and terrorising your students. Two have been wrongfully suspended, and my son and his friend were coerced into lying to police. Mayfield High has a serious crisis on its hands, and we expect it to be treated as such. We will be coming in tomorrow with our lawyer."

The word lawyer changed everything. The school was no longer dealing with a troubled student. It was facing potential legal consequences.

The most difficult and most necessary intervention came that evening. After a tense phone call between the families, Leroy and Danielle agreed to meet with their parents present. They gathered in the Millers' living room, the air thick with awkward, painful silence.

Leroy and Danielle sat stiffly on the edge of the sofa, their posture defensive, their eyes filled with exhausted, justified anger. Their parents sat close, a quiet wall of protection.

Kyle, seated between his own parents, knew this was his responsibility. He drew a deep, shuddering breath and looked directly at his former friends.

"I don't know where to start," he said, his voice raw. "And 'sorry' isn't big enough for what I did. I didn't just let you down. I participated. I chose the game over you, over our friendship, because I was a coward. I was so terrified of my secrets being exposed that I stood by while it destroyed you. And then… I helped it."

He looked at Leroy. "The things it made you look like, a vandal, a criminal. I knew it was a lie. And I said nothing."

He turned to Danielle. "And you… I let it turn the whole school against you. The things I said were monstrous. There's no excuse. I was a terrible friend, and I will spend the rest of my life trying to make up for it."

It was not a polished apology. It was raw and painful, and the room fell silent as he finished.

Danielle was the first to speak, her voice quiet but steady. "Why, Kyle? Why did you keep going, even after you saw what it was doing to us?"

"Because I was trapped," Kyle whispered, the words sounding fragile even to his own ears. "It had a file on me. On all of us. It threatened to release it if I didn't comply. But that's not an excuse. It was still my choice. And I chose wrong. Every single time."

It was Danielle's mother who broke the tension. "This 'Glitch,'" she said, her voice sharp with anger. "The police need to be involved. This is a criminal matter. This is targeted harassment and coercion of minors."

The conversation shifted from personal betrayal to collective strategy. The four parents, united by a common and invisible enemy, began to talk. They discussed lawyers, police reports, and a coordinated approach to force the school into taking real, systemic action.

As the adults spoke, Kyle caught a flicker of something in Leroy's eyes. It was not forgiveness, not yet. It was recognition. The empty chair at their table had once symbolised their broken friendship, but now, sitting in the same room, they were all victims of the same predator.

The intervention was not about instant forgiveness or a neat resolution. It was about laying the full, ugly truth on the table and forming a difficult but necessary alliance from the wreckage. They were no longer four teenagers trapped in a secret game. They were four families, armed with the truth and preparing to fight back. The power of The Glitch had been its ability to isolate them. Their first and strongest act of defiance was to stand together.

Chapter 29:
Rebuilding the Code

The fallout from Kyle's public confession hit Mayfield High like a tsunami, leaving no corner of the school untouched. The chaos of the assembly gave way to a week of frantic meetings, parental outrage, and a slow, painful reckoning.

Faced with a united group of parents and the looming threat of legal action, the school was forced into an immediate and unprecedented response. The flimsy narrative of "a few troubled students" was abandoned. In a second, far more sombre assembly, Mr Harrison addressed the entire school, speaking openly about the dangers of a malicious, predatory piece of software that had targeted members of their community.

The school's Rebuilding the Code initiative was swift and wide-ranging. A campus-wide ban on mobile phone use during school hours was introduced. A mandatory digital citizenship program was rolled out for all year levels. Most importantly, the suspensions of Leroy and Danielle were publicly overturned, with a formal written apology issued to their families for the school's failure to investigate the situation properly.

For Kyle, school became a strange mix of notoriety and isolation. He was no longer invisible, but he was a pariah to some and a curiosity to others. He was the boy who had collapsed on stage, the one who had exposed the darker side of their digital lives. He walked the corridors to a constant murmur of whispers, but he no longer cared. Fear of public opinion was now a distant echo, drowned out by the harder work of rebuilding himself.

His recovery began with therapy. His sessions with Dr Anya Sharma, the trauma specialist, were nothing like the dismissive meetings he had endured with the school counsellor. Dr Sharma listened without judgement as Kyle carefully retraced his descent into The Glitch. She gave him language for what had happened, terms like coercive control, gamified manipulation, and trauma bonding.

"You weren't addicted to a game, Kyle," she explained during one session. "You were being held hostage. The app created a problem, the threat of exposure, and then offered you the solution, which was compliance. It's a classic psychological trap. Your choices were made under pressure."

The sessions were exhausting. He was forced to face his shame, to separate the avatar's cold logic from his own fear, and to accept that vulnerability did not make him weak. It made him human.

His relationship with Aaron was the most fragile part of his recovery. Their shared trauma bound them together, but Kyle's betrayals had left deep, festering wounds. At first, they met in the neutral, clinical safety of Dr Sharma's waiting room, their parents having coordinated appointments.

Their first conversation was stiff and awkward. Aaron's arm was out of the cast, but he still held it carefully, a physical reminder of the lie that had shaped their lives for months.

"I'm sorry, Aaron," Kyle said, the words feeling small and inadequate. "For everything. For not listening. For using your accident in the assembly. For being a terrible friend."

Aaron did not look at him. He stared at the worn carpet, his foot tapping a nervous rhythm. "You were scared," he said, not as an excuse, but as a statement of fact. "I was too. But you changed. You became like it. Cold."

"I know," Kyle whispered. "I felt it happening. It was like someone else was inside my head, making the 'logical' choice every time. And I let it."

It was the start of a slow and difficult thaw. Their friendship did not snap back into place. It was damaged, and rebuilding it would be quiet and painstaking, shaped by shared therapy sessions and the gradual return of trust. They began talking again, not about The Glitch, but about ordinary things, music, school, the ridiculousness of the new phone ban. It was tense, but it was a beginning. The friendship was salvageable, though it would always carry scars.

His relationship with Leroy and Danielle was even more complicated. While official apologies had been made, personal betrayal could not be repaired by a school initiative. He passed them in the corridors and received a brief, hesitant nod, an acknowledgement of shared history, but the easy camaraderie was gone, replaced by guarded distance. Kyle knew rebuilding that bridge would take months, perhaps years, and he accepted it as a consequence of his choices.

The digital world, once his refuge, now filled him with anxiety. He was given a new phone, a basic model with no social media apps, and he used it sparingly. Phantom vibrations still came from time to time, a lingering form of digital trauma, but they were fading. He was learning to live in the real world again, to find meaning in the slow, unrewarded rhythms of everyday life. He was rebuilding his own code, one painful, honest line at a time.

Chapter 30:
Facing the Glass

Dr Anya Sharma's office was nothing like the sterile rooms at school. It was warm, filled with softly lit lamps and overflowing bookshelves, and it smelled faintly of old paper and chamomile tea. It was a safe harbour, though for Kyle, the early sessions felt like navigating a storm. He was used to hiding, deflecting, performing. Honesty was a language he was relearning from the ground up.

"The Glitch didn't create your fears, Kyle," Dr Sharma said gently but firmly. They were in their third session, and she was guiding him beyond a simple recounting of events and into the dangerous territory of his inner world. "It found them. It found the fault lines that were already there. Can you tell me about the boy in that diary entry? The one who felt like a fraud?"

The question landed hard. He shifted in the armchair, the familiar knot of shame tightening in his stomach.

"He was scared," Kyle admitted quietly. "Scared of not being good enough. At football, at school, at anything. It felt like everyone expected me to be the smart, reliable kid. And it felt like a lie. I always felt one mistake away from everyone finding out I was just… average."

"And what's so wrong with being average?" she asked, her gaze steady and perceptive.

The question stunned him. In the world of The Glitch, average meant failure. Status, rank, being Elite, those were the only measures of worth. The idea that being simply okay might be enough was something he had not considered in a very long time.

"In the game, it's everything," he muttered. "If you're not at the top, you're nothing. You're a liability."

"But you are not in the game anymore," Dr Sharma said softly. "You're here. In this room. And here, you don't need to be Elite. You just need to be honest."

Through these careful, difficult conversations, Kyle began to understand his vulnerability. The Glitch had not only exploited his fear of failure; it had offered him an escape. A world with clear rules, measurable success, and the illusion of control, without the uncertainty of real life. He had to face the uncomfortable truth that part of him had wanted that escape, had craved the validation the app offered so easily. The guilt over his choices remained heavy, a weight he carried every day, but for the first time, he was learning how to set it down.

The hardest part of his recovery, however, was not in the therapist's office. It was in the hallways of Mayfield High. The school, in its attempt to rebuild the code, had organised a series of voluntary reconciliation sessions, mediated by Mrs Gable, the school counsellor. Kyle knew he had to attend, not for the school, but for himself. He needed to face Leroy and Danielle.

He found them waiting in a small, empty classroom. The air was heavy with awkward, painful tension. They sat on opposite sides of a table, the empty space between them a clear sign of their fractured friendship.

Kyle had rehearsed what he wanted to say a hundred times, but the words felt clumsy and inadequate as they left his mouth.

"I know I've said sorry before," he began, his voice shaking slightly. "But I need to say it again. To your faces. What I did, helping the app to hurt you, staying silent, choosing it over you, it was the worst thing I've ever done in my life. And I know you have no reason to ever trust me again. I just needed you to hear it from me. I was wrong."

Leroy, who had been staring at the table, finally looked up. His eyes, once full of quick, cynical humour, were now guarded and tired.

"Do you know what the worst part was, Kyle?" he asked, his voice flat. "It wasn't the suspension. It wasn't the rumours. It was knowing that you knew the truth and still chose to bury us. I kept waiting for you to step in, to defend us. And you never did."

"I was a coward," Kyle admitted, the words tasting like poison. "I was so trapped by my own fear that I couldn't see past it. It's not an excuse, but it's the truth."

Danielle, who had been silent, finally spoke, her voice quiet but carrying undeniable strength. "The app used our weaknesses against us. It used your fear of being average. It used Leroy's cynicism. It used my loyalty to my family." She looked at Kyle, her gaze steady. "It turned our best qualities into weapons. But we all had a choice. Leroy and I chose to walk away from that bridge. You chose to cross it."

Her words were a stark and painful summary of their journey. There was no accusation in her tone, only a statement of fact. The apology was offered, and it was heard, but it was not a magic fix. It could not erase the choices that had been made.

"I know," Kyle said softly. "And I have to live with that."

The conversation did not end with a hug or a promise to be friends again. It ended with a quiet, hesitant nod. It was an acknowledgement of shared trauma, a tentative first step on a long and uncertain road. Trust, Kyle realised, was not something that could be restored with a single apology. It had to be rebuilt, piece by painful piece, through consistent and honest action over time.

He left the classroom feeling emotionally drained, but also strangely lighter. He had faced the glass, not the screen of a phone, but the reflective and honest gaze of the friends he had betrayed. He had shown them his guilt and his shame, and they had not turned away. The trust was not there yet, but for the first time, he felt a flicker of hope that one day it might be.

Tara Quinn

Epilogue
One Year Later

The late afternoon sun cast long, golden stripes across the football pitch. Kyle took a deep breath, the familiar earthy smell of grass and damp soil filling his lungs. He was playing again, not with the frantic need to win that had once driven him, but with a quiet enjoyment of the game itself. The final whistle blew, signalling a hard-fought draw. He clapped his teammates on the back and shared a laugh with the captain. His life was, for the most part, back to normal. But it was a new kind of normal, one that had been hard-earned and felt far more precious.

His grades were steady, his focus had returned, and the hollow, haunted look in his eyes had been replaced by a calmer, more thoughtful gaze. He was quieter now, less concerned with status and more aware of the subtle, unspoken currents of the world around him. He had stood at the edge of a digital abyss and, by some miracle, found his way back.

The scars remained. They were invisible, but he felt them every day. He still used a basic phone, and sometimes a stray pop-up ad or an unexpected system update notification sent a rush of cold adrenaline through his body, a brief and paranoid fear that it was back. He knew the pull of the digital world, and the easy validation of a screen, was not a battle he had won forever, but a lifelong commitment to awareness and restraint.

The Glitch app itself had vanished. After the school assembly, media attention and a formal police investigation led to a nationwide cyber alert. The app, and the anonymous servers that hosted it, disappeared from the web as quietly as they had arrived. It was untraceable, a digital ghost. But the idea of it lingered. Kyle saw its shadow in the addictive design of new social media platforms, in the obsessive way his peers curated their online lives, and in every app that promised status and connection for the low price of time and personal data.

His relationship with his parents was stronger and more honest than it had ever been. The crisis had broken their old dynamic, and in its place they had built something new, grounded in painful but genuine understanding.

He and Megan were closer too. She had forgiven his distance, and now their sibling banter carried an unspoken layer of protective affection.

He and Aaron were still best friends, but their friendship had been reshaped by everything they had faced together. It was deeper and quieter, marked by a shared understanding of the darkness they had survived. They did not talk about The Glitch, but they did not need to. Some things were understood without words.

His connection with Leroy and Danielle was rebuilding slowly and carefully. The anger had faded, replaced by a cautious and respectful distance. They spoke occasionally, sharing a quiet joke in the hallway. The trust was not what it once had been, but a fragile new foundation was taking shape, built on the shared experience of surviving the same storm.

The previous week, as part of the school's now permanent digital wellbeing program, Kyle had volunteered to speak to a group of Year 9 students. He did not tell them the full story. Instead, he shared a version of it, speaking quietly about the lure of online status, the danger of trading real connections for digital rewards, and the importance of knowing when to log off. He used his experience not as a weapon, but as a warning, a quiet signal for any other kid who might be tempted by a glitch in their own perfectly ordinary life.

He walked home from the football match, his bag slung over his shoulder, the evening air cool against his skin. He pulled out his phone, checked a single text from his mum asking what time he would be home, and put it away. He did not scroll. He did not check notifications. He did not need the validation.

He paused at the end of his street, taking in the familiar and comforting sight of his house, its windows glowing with warm light.

Kyle looked at the world around him, the cracked pavement, the tangled branches of the trees, and the solid presence of home. He knew, with a certainty that settled deep within him, that you cannot log off from reality.

And for the first time in a very long while, he did not want to.

Resources & Support

Thank you for joining Kyle on his journey. The themes in this book, such as digital addiction, cyberbullying, and mental health struggles, are intense, but they are also real. If you or someone you know is facing challenges like the ones shown in this story, please know that you are not alone and that help is available.

Reaching out is an act of strength. Below are some organisations in the UK that offer free, confidential support and information.

For Mental Health and Emotional Support

If you are struggling with your mental health, feeling overwhelmed, or simply need someone to talk to:

Young Minds

A leading UK charity committed to improving the mental health of children and young people. They offer information, resources, and a parents' helpline.

Website: youngminds.org.uk

Text support: Text YM to 85258 for free, 24/7 support.

The Mix

Provides essential support for under 25s on a wide range of issues, from mental health to money worries and online safety. They offer a helpline, webchat, and counselling services.

Website: themix.org.uk

Helpline: 0808 808 4994

Mind

A leading mental health charity in England and Wales. They provide advice and support to empower anyone experiencing a mental health problem.

Website: mind.org.uk

Infoline: 0300 123 3393

For Bullying and Online Safety

If you are experiencing cyberbullying or are concerned about your safety online:

The National Bullying Helpline

Offers practical advice and support to anyone affected by bullying, whether at school, online, or in the community.

Website: nationalbullyinghelpline.co.uk

Helpline: 0300 323 0169

Childnet International

A non-profit organisation working with others to help make the internet a safe and positive place for children and young people. They provide excellent resources on online safety.

Website: childnet.com

For Immediate Crisis Support

If you are in crisis and need to speak to someone urgently:

Childline

A free, private, and confidential service where you can talk about anything. Whatever your worry, whenever you need help, they are there for you online and on the phone.

Website: childline.org.uk

Helpline: 0800 1111, available 24/7

Samaritans

Available 24 hours a day, 365 days a year. If you need a safe place to talk, at any time and in your own way, they are there to listen.

Website: samaritans.org

Helpline: 116 123, free from any phone

Remember, your story matters, and so does your wellbeing. Please take care of yourself.

International Support

Mental health and online safety are global issues. If you are reading this outside the UK, the organisations below can offer support in your region.

For Readers in the United States

The Trevor Project

The world's largest suicide prevention and crisis intervention organisation for LGBTQ young people.

Website: thetrevorproject.org

24/7 crisis line: 1 866 488 7386 or text START to 678678

Crisis Text Line

Provides free, 24/7 text-based mental health support and crisis intervention through trained volunteers.

Website: crisistextline.org

Text support: Text HOME to 741741

The National Suicide Prevention Lifeline

A national network of local crisis centres providing free and confidential emotional support, 24 hours a day, 7 days a week.

Call or text: 988

For Readers in Australia

Kids Helpline

Australia's only free, private, and confidential 24/7 phone and online counselling service for young people aged 5 to 25.

Website: kidshelpline.com.au

Helpline: 1800 55 1800

Headspace

The National Youth Mental Health Foundation, providing early intervention mental health services for 12- to 25-year-olds and promoting young people's wellbeing.

Website: headspace.org.au

eSafety Commissioner

Australia's national agency for online safety and a key resource for reporting cyberbullying and seeking support for negative online experiences.

Website: esafety.gov.au

For Readers in Europe

Mental Health Europe

Represents organisations and individuals working in the field of mental health and wellbeing across Europe. While not a direct helpline, their website is a useful starting point for finding country-specific resources.

Website: mhe-sme.org

European Emergency Number

If you are in immediate danger anywhere in the European Union, call 112. This universal emergency number connects you to local police, ambulance, or fire services.

A Note for All Readers

The resources listed here are a starting point. Many countries have their own dedicated mental health and online safety support services. Searching for "youth mental health helpline" or "cyberbullying support" along with your country name will often help you find local options.

Reaching out is the first and most important step. You are not alone.

Thank You for Reading

If these characters and their journey have touched your heart, please consider leaving a review. Reviews are small echoes of kindness for an author. They help new readers discover the story and mean more than you might realise. A few thoughtful words on the site where you purchased this book can make a real difference.

About the Author

Tara Quinn was born in Ireland, a land where myths feel like memories and stories are woven into the landscape itself. This early immersion in folklore sparked a lifelong passion for storytelling, which she carried into a career in education spanning more than fifteen years.

Working closely with children from all walks of life, Tara recognised a universal need for stories that act as both mirrors and windows. Mirrors that reflect a child's own experiences, and windows that open onto a wider, more diverse world, building empathy and understanding.

Her work in the classroom solidified her mission to create learning opportunities that go beyond textbooks. She is a passionate advocate for using storytelling to foster curiosity, encourage academic learning, and nurture essential social and emotional skills, helping children gain the knowledge and confidence to navigate their own complex inner worlds.

www.ingramcontent.com/pod-product-compliance
Lightning Source LLC
Chambersburg PA
CBHW031303060726
47590CB00003B/1046